MEMBRANE-MEDIATED INFORMATION:
VOL. 1
BIOCHEMICAL FUNCTIONS

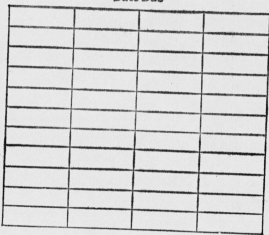

MEMBRANE-MEDIATED INFORMATION: VOL. 1 BIOCHEMICAL FUNCTIONS

Edited by

P. W. Kent

Director of the Glycoprotein Research Unit and
Master of Van Mildert College
Durham University
England.

AMERICAN ELSEVIER PUBLISHING CO., INC.
NEW YORK

Published in the United States by
AMERICAN ELSEVIER PUBLISHING CO. INC.
NEW YORK

Published in the United Kingdom by
MTP
MEDICAL AND TECHNICAL PUBLISHING CO. LTD.
LANCASTER

Library of Congress Catalog Card Number 73–17994

ISBN:0–444–19540–8

First published 1973

Printed in Great Britain

Contents

The late R. J. Winzler–An Appreciation
Preface
Acknowledgements

The Late Richard J. Winzler; An Appreciation

On his return home, and within a few days of completing his contribution to this book, Richard Winzler died quite suddenly and unexpectedly.

There was taken from us not only an able and dedicated scientist who had been a veritable pioneer in the biochemistry of glycoproteins but also a most lovable person whose wide circle of friends in all parts of the world is a tribute to this genial and humane personality. The following words were spoken at his funeral on the 29th September, 1972, by Dr. Earl Frieden, Chairman of the Chemistry Department, Florida State University, where Richard and his group worked.

"Dick Winzler would have been 59 years old today. In those 59 years he had crowded 100 years of living—it was beautiful living, interwoven with his wife Janna and their children, Joan, Natalie and Lee, and 35 years as an outstanding biochemist and a dedicated teacher of chemistry and the medical sciences. His life started in San Fransisco, progressed through his Bachelor's and Ph.D. degree at Stanford, the latter at the remarkably young age of 24. This was followed by a series of coveted fellowships at Yale, at the Wenner Grens Institute in Stockholm, at Cornell and at the National Cancer Institute, leading to his first faculty position in the Department of Biochemistry at University of South California in 1943. It was there I met him in 1944 and was immediately captivated by this sparkling, bubbling, irrepressible and imaginative young Assistant Professor.

And with his generous help and guidance, I became the second graduate student to complete the Ph.D. under his supervision. This brief statement does not do justice to the great support, both moral and practical, which Dick Winzler gave me and his many other students during our days as graduate students. Many of his students have gone on to major achievements in the universities and research laboratories of this nation. For me it was the start of a lifelong warm, personal friendship as well as a maturing professional association. Over the years our relationship evolved from an intellectual father-son to a much more brotherly one.

From a professorship at University of South California he moved to the University of Illinois, College of Medicine, where he became Head of their large Department of Biochemistry; 13 years later, he became Head of Biochemistry at State University of New York in Buffalo. It was mainly during this 15–20 year period that Dick Winzler published most of his 168 research papers and achieved international recognition from his favourite research areas: in glycoprotein and membrane chemistry, particularly in their relation to cancer cells. He was appointed to serve on several editorial boards, e.g., Cancer Research, and to the Board of Directors of numerous important national policy committees of the American Cancer Society and the National Institute of Health.

In 1969 we persuaded him to join Florida State University under our National Science Foundation Science Development program. His years here were developing into his most productive period. Undisturbed by chairmanship responsibilities, his love for academic life achieved its full fruition here. His ability to bridge the generation gap, to communicate effectively, made him an outstanding teacher. His rich background in medical education was a great asset to Florida State University's burgeoning medical program. But what impressed us most about Dick Winzler was his boundless zeal for living and learning. If I were to try to describe him in a few words, they would be: energetic, intelligent and compassionate. He was like a molecule of ATP; he was our high energy intermediate. He had an unusually high group transfer potential—he was able to transfer his enthusiasm for teaching and research in biochemistry to all those around him.

At a time like this, we tend to be overwhelmed by our loss. I certainly feel that I have lost a brother as well as a most valued colleague. But we must not forget what we have gained by knowing and working with Dick Winzler, the inspiration of his energy and his zest for biochemistry, and the memory of his friendship, his special bounce and his joy for living".

Preface

In the past two decades, the attention of biochemists has been almost entirely directed towards nucleic acids and proteins as the macromolecules which convey biological information. Indeed, it has been held that these alone could have such a function. The dramatic progress in molecular biology and the consequent understanding of the molecular basis for the transmission of genetic information and of its implementation in cell function has done much to reinforce this view. The formulation of the Central Dogma in 1957 set down principals for the flow of information from the fundamental deposit held in the DNA-codes of each cell nucleus to the process of protein biosynthesis by means of which distinctive enzymic functions are implemented. *Once "information" has passed into protein it cannot get out again.* The passage of information from protein to protein or from protein to nucleic acid as thus envisaged is forbidden. The theory as propounded is primarily concerned with proteins as enzymes and with the enzyme specific activities arising from unique amino acid sequences and conformations. The cell is thus genetically inviolate from its environment save in so far as the potential information in its nuclear store allows and is capable of implementation.

Since that time, ongoing biochemical research especially in immunology and cytology, has shown that types of specific biological information exists other than DNA- and RNA-based codons. In particular, the mechanism of molecular recognition apparently implicit in antigen–antibody interactions and in a variety of cell-recognition phenomena associated with cell membranes point to the possibility of other forms of information being conveyed through macromolecules. Detailed investigation of the structure of the determined groups conveying blood-group specificity of the A, B and H system has shown that both glycoproteins and glycolipids possess these functions. The acuity of the information is derived from a series of specific oligosaccharides attached to a protein chain at the time of or after its biosynthesis as an ordered sequence. Each informational oligosaccharide requires to be complete in certain essential structural details if it is to perform its function, and if membrane-bound, probably to be attached in some definitive topographical way. This volume is concerned with an exploration of further possibilities of membrane-mediated information, its basis and regulation.

Though '*biologists should not deceive themselves with the thought that some new class of biological molecules, of comparable importance to the proteins remains to be discovered*' there is now the very real likelihood that the informational content of proteins is not exhausted by the amino acid sequences. Post-polymerization modification, such as glycosylation, phosphorylation, sulphation or possibly deamidation offer significant and cogent means for adding distinctive information to the protein. It may be argued that such informational overtones must necessarily be of secondary significance since these also can only be brought about by the action of specific enzymes, themselves arising from the genetic store. However, "information" of this sort differs from the sequence-derived informational content of protein in that it appears to be susceptible to modification at the enzyme level (e.g. by kinetic and specificity effects) without recourse to the genome. A number of reports in this volume are concerned with this phenomenon. The final molecular structure of a glycoprotein and possibly of glycolipids also (as far as the carbohydrate moiety is concerned) would thus appear to be dependent on the metabolic state of the cell and its environment. The implications of these possibilities for biology are considerable and outline further possible mechanisms by which biochemical individuality can be defined, as well as further degrees of flexibility, enabling cells to respond to limited changes required by their surroundings.

P. W. Kent

Acknowledgements

This, the first of two volumes, stems from discussions of an international and interdisciplinary nature which took place at Christ Church, Oxford, in September, 1972, under the auspices of the Board of Management for the Foster and Wills Scholarships at Oxford University and of the German Academic Exchange Service.

These were made possible by generous financial support of the Kulturabteilung of the German Foreign Office, the Stifterverband für die Deutsche Wissenschaft and the Deutscher Academischer Austauschdienst which is most gratefully acknowledged. Thanks are also due to the Governing Body of Christ Church for its part in making facilities available, and in particular to the Steward, Mr. E. M. James. The notable contributions, made to the organizational and secretarial aspects of the work by Mr. M. L. Mruck, his staff, and Mr. N. Gascoyne are also warmly acknowledged.

The Foster and Wills Scholarships exist to support the exchange each year of young scholars in any field and of either sex, between Oxford and German Universities. The contributors to this book have agreed unanimously that any proceeds arising from its sale shall be given to that Scholarship fund.

The editor thanks the respective authors and owners of the copyrights of quoted material for their generous cooperation in allowing us to reproduce items from their publications.

Dedicated to the memory of
R. J. Winzler
with Affection and Respect

I Carbohydrates of Cell Surfaces

1 A Model for Control of Oligosaccharide Structure in Membrane Glycoproteins

The late R. J. Winzler

Department of Chemistry, Florida State University

Marked variations in the ratios of sialic acid, fucose, and sulphate were observed in the mucins from submaxillary secretions of the dog depending upon the stimulus inducing secretion. The aminoacid composition of the mucins, however, was indistinguishable. This suggests that the nature of the oligosaccharides of secreted mucin is under some type of physiological control. Since such variation could also be involved in regulating the structure of membrane glycoproteins, and thus of membrane-mediated information, a detailed study was undertaken of the structures of oligosaccharides in canine submaxillary secretions. Two types of acidic oligosaccharide chain were observed, one containing sialic acid and no sulphate, and the other sulphate but no sialic acid. It was postulated that both types of oligosaccharide chain are on the same protein, but that their proportions can be quite variable depending on the physiological state of the secreting submaxillary gland. It is proposed that biosynthesis of one or the other of the oligosaccharide types is determined at the disaccharide stage when either a sialic acid is added to N-acetylgalactosamine, or N-acetylglucosamine is added to galactose in the growing oligosaccharide chain.

The plasma membranes of cells are known to carry significant amounts of carbohydrate on their external surfaces (Martinez-Palomo, 1970; Rambourg, 1972; Winzler, 1970, 1972). In certain instances this carbohydrate is clearly involved in reactions of cells with exogenous molecules, e.g. the interaction of the A, B, H and Lewis antigens on human erythrocytes with their respective antibodies. It is tempting to speculate that such recognition by, and specific interaction with, exogenous molecules may be a major role of these carbohydrates on cell surfaces. This chapter involves such speculation, and explores a possible model by which the nature of the oligosaccharide chains of glycoproteins may be modified by physiological factors.

CARBOHYDRATES AS INFORMATIONAL MOLECULES

Macromolecules containing carbohydrates have immense potential for storing information in small molecular units. The individual monosaccharides in their pyranose or furanose rings are relatively rigid molecules with hydroxyl groups capable of forming hydrogen bonds oriented in very fixed steric relationships. In addition the linkages between carbon 1 of the non-reducing terminal sugar and the penultimate sugar may be at the 2, 3, 4 or 6 hydroxyl groups. This glycosidic linkage may have either an alpha or a beta configuration. Each of the eight possible disaccharides resulting from a given pair of monosaccharides will have a characteristic shape and hydrogen bonding potential. It is no wonder, then, that carbohydrates have tremendous potential as informational molecules. Figure 1.1 compares the potential forms of a terminal, non-reducing disaccharide and an amino terminal dipeptide, and shows

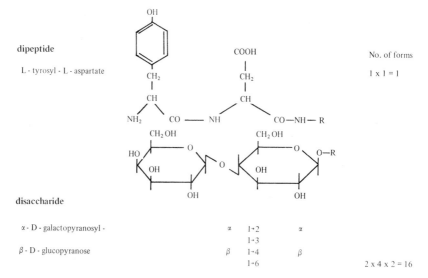

Fig. 1.1 Potential versatility of carbohydrates for cell–cell recognition and other information processes.

that there are 16 possibilities for the structure of the disaccharide and only one for the dipeptide.

Microheterogeneity

Glycoproteins occurring in membranes, tissues and body fluids generally show a microheterogeniety dependent upon differences in their carbohydrate components. There may be several types of such microheterogeneity. One type is genetically controlled by the specificity of glycosyl transferases governed by allelic genes. It is very likely, for example that the A, B and Lewis blood-group glycoproteins have identical peptide chains, but differ only in the outer members of the oligosaccharide chains which give them their characteristic antigenic activity (Watkins, 1972). The same is likely to be true of the M and N antigens, and perhaps many others as well. In the case of the AB individual, the peptide core probably contains both A and B antigens, possibly in variable ratios, and the glycoprotein, therefore, would exhibit microheterogeneity due to differences in the carbohydrate components.

Microheterogeneity in glycoproteins not so directly related to the genetic makeup of the individual, also may occur. Thus, every glycoprotein that has been carefully studied has shown microheterogeneity due to incompleted or missing chains (Cunningham, 1971; Montgomery, 1972). Such microheterogeneity has been demonstrated in the major glycoprotein of the human erythrocyte membrane (Thomas and Winzler, 1969, 1971) as well as in plasma α_2-macroglobulin (Dunn and Spiro, 1967), in plasma orosomucoid (Wagh *et al.*, 1969) and others. Even the ovalbumin isolated from the eggs of a single chicken has shown this sort of microheterogeneity, and has yielded glycopeptides with differing ratios of mannose and acetyl glucosamine. It is difficult at this time to know whether this kind of microheterogeneity has biological significance. However, the possibility exists that incompleted oligosaccharides may be involved in cell–cell or cell–protein attachment by virtue of their interactions with membrane-bound glycosyl transferases forming 'enzyme–substrate complexes' to give reversible attachment of the cells to other cells or to glycoproteins (Roseman, 1970; Barber and Jamieson, 1971).

What appears to be a third type of microheterogeneity of glycoproteins, one which may be under physiological control, was suggested by the work of Dische and his collaborators in 1962 (Dische *et al.*, 1962; Dische, 1963). In these experiments the submaxillary gland of the dog was cannulated, and submaxillary secretion was elicited either by electrical stimulation of the Ramus

communicans nerve, or by intravenous administration of pilocarpine at various levels. A relatively crude mucin was isolated from
the secretions, and the amounts of neutral sugar, fucose, amino
sugars and sialic acid were determined in the isolated mucins. It was
observed that a reciprocal relation existed between sialic acid and
fucose, the molar ratios ranging from 0.2 to 2.0. However the sum
of the two remained the same. This is shown in Table 1.1.

Table 1.1
COMPLEMENTARY RELATIONSHIP BETWEEN
SIALIC ACID (SA) AND FUCOSE (FUC) IN
CANINE SUBMAXILLARY MUCIN
(Extremes from Dische et al., 1962)

	SA:FUC	$\dfrac{\text{SA + FUC}}{\text{HEXOSAMINE}}$
Pilocarpine (0.4 mg)	2.1	1.2
Pilocarpine (4 mg)	0.2	1.1
Electrical	0.3	1.1

Sialic acid and fucose are terminal, non-reducing sugars in
most mammalian glycoproteins, and Dische suggested that they
might compete to terminate a growing oligosaccharide chain, with
the competition being in some way under physiological control.
Dische also pointed out that the physical properties of the sialic
acid-rich mucin should be different from those of the fucose-rich
mucin, since sialic acid contains a carboxyl group whereas fucose
does not.

We became interested in these observations since they suggested
that a rather novel control mechanism might be operative in mucin
biosynthesis. The data suggested that, depending upon the physiological state, there could be rapid activation or inhibition of specific
sialic acid or fucose transferases, or that there could be a change in
the pools of CMP-sialic acid or GDP-fucose, or that there could be a
selective stimulation of the biosynthesis of different glycoproteins.
Such physiological control of the chemistry of secreted mucins may
well have a counterpart in the control of the chemistry of membrane
glycoproteins, some of which resemble mucins in their chemical
structures.

I should like to report observations made in collaboration with
Drs E. A. Johnson and C. Lombart, which lead us to conclude that
there are two types of acid oligosaccharides in canine submaxillary
mucin, and that the ratio of the two types can be changed by physiological manipulation.

Physiological Control of Carbohydrate Composition of Canine Submaxillary Mucin

Canine submaxillary mucin was isolated from frozen sub-maxillary glands by water extraction, chromatography on Sephadex G-200, and removal of contaminating proteins by chromatography on biogel CM 100 at pH 4.5 (Lombart and Winzler, 1972).

Mucin was also isolated from submaxillary secretions by essentially the same method, or by precipitation with cetyl trimethyl ammonium bromide. Submaxillary ducts were cannulated and secretion was elicited in the dogs either by electrical stimulation of the Ramus communicans nerve or by the intravenous infusion of pilocarpine. The electrical stimulation and the infusion were controlled so that the submaxillary secretion rate was close to 1 ml $\times$ min^{-1} under both conditions.

From 1 to 3 mg of purified mucin was isolated from each millilitre of submaxillary secretion. Neutral sugar was determined in each sample by the phenol sulphuric acid method of Dubois *et al.* (1956) and the individual neutral sugars by gas liquid chromatography of the alditol acetates (Lehnhardt and Winzler, 1968). Sialic acid determination was by the method of Aminoff (1961). Amino sugars and their alcohols were quantitated by the ion exchange method of Weber and Winzler (1969).

We could confirm Dische's observations on the reciprocal relation between sialic acid and fucose in the secreted mucins. However when we attempted to demonstrate differences in the electrophoretic mobilities of mucins having high and low sialic acid : fucose ratios, no significant difference was found. We therefore suspected that sulphate might be contributing to the negative charges of the mucin (Bignardi *et al.*, 1964), and so determined the sulphate content of the mucin using the method of Antonopoulos (1962).

Figure 1.2 shows the carbohydrate and sulphate content of mucin isolated from successive 50 ml samples of submaxillary secretion from a dog in which one gland was stimulated electrically for 200 min and then the other by pilocarpine administration for 150 min. The composition of the mucin collected over a 200 min period as a result of electrical stimulation remained relatively constant, with a molar ratio of sialic acid to fucose of about 0.5 and a ratio of sialic acid to sulphate of about 0.75. However when secretion from the contralateral gland was elicited by pilocarpine stimulation, the initial composition of the mucin was very different from that obtained from electrical stimulation. Specifically it was higher in sialic acid, and lower in fucose and sulphate (SA:Fuc = 3.3, SA:SO$_4$ = 4.7). Over the next 150 min the concentrations of sialic

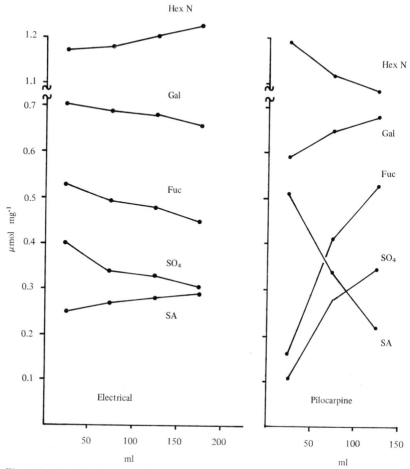

Fig. 1.2 Carbohydrate and sulphate content of mucin secreted by dog sub-maxillary gland, differently stimulated.

acid, fucose and sulphate approached those obtained with the electrically-stimulated material. This experiment was repeated many times, always with the same results, i.e. the early pilocarpine secretions yielded mucins with high sialic acid : fucose and sialic : sulphate ratios which gradually fell to the levels observed in the stored mucin and in the mucin elicited by electrical stimulation. In all cases sialic acid plus fucose and sialic acid plus sulphate remained essentially the same. These relationships are summarized in Table 1.2. Since the sum of sulphate and sialic acid was constant, the charge on the mucin was not changed during the transition.

It appeared possible that the mucin secreted immediately following pilocarpine administration might represent glycoprotein

Table 1.2
CARBOHYDRATES IN CANINE SUBMAXILLARY MUCIN

	From Gland	Secretion (Elec.)	Secretion (e Pilo)	Secretion (l pilo)
		μmol/100 g		
Galactose	81	82	64	83
Fucose (F)	63	66	20	68
NANA	28	30	58	28
GalNAc	69	56	56	46
GluNAc	56	58	54	46
SO_4	—	47	22	49
NANA/F	0.44	0.45	2.9	0.41
NANA + F	0.91	0.96	0.80	0.78
NANA/SO_4	—	0.64	2.6	0.57
NANA + SO_4	—	0.77	0.80	0.87

(e = early, l = late)

different from that secreted under electrical stimulation. In fact
Pigman and co-workers have generally noted a major mucin and a
minor mucin in extracts of submaxillary glands (Tettamanti and
Pigman, 1968). We made many unsuccessful efforts to fractionate
the early and late pilocarpine mucin to obtain sialic acid-rich and
fucose-rich fractions. In spite of the marked variation in the sialic
acid, fucose and sulphate content of the secreted mucins, it was
observed that the amino acid composition was similar or identical
in all of mucins studied. This is shown in Table 1.3 which shows
moles of amino acid per 100 residues in mucin isolated from the
gland, from electrically-stimulated mucin, and from early and late
pilocarpine secretions. No significant difference in the amino acid
composition of the different mucins is evident. Although similarity in
amino acid composition does not establish that the mucins isolated
have a single type of peptide core with variable oligosaccharide
chains, it is our feeling that this is indeed the case.

Acidic Oligosaccharides in Canine Submaxillary Mucin

Our next efforts were directed at determining the structures of
the oligosaccharide chains of the secreted canine submaxillary
mucin, and to ascertain whether there were differences in the nature
of these chains depending upon the nature of the stimulus inducing
secretion. Mucin labelled with radioactive sulphate was prepared by
infusing $^{35}SO_4^{2-}$ into the facial artery. After one hour, secretion was
induced by electrical stimulation. The labelled mucin was isolated
and subjected to treatment with alkaline borohydride (0.05 M

Table 1.3
AMINO ACIDS IN CANINE SUBMAXILLARY MUCIN

	mol/100 mol			
	From Gland	Secretion (Elec.)	Secretion (e Pilo)	Secretion (l Pilo)
Lysine	1.4	1.4	1.6	1.5
Histidine	0.8	0.6	0.8	0.7
Arginine	4.0	3.8	3.9	4.1
Aspartic	3.4	3.5	3.9	3.9
Threonine	13.0	12.1	13.0	12.4
Serine	10.0	10.1	10.9	10.6
Glutamic	7.4	7.7	7.4	7.0
Proline	11.5	11.8	13.2	12.8
Glycine	23.5	23.7	21.2	23.9
Alanine	11.3	11.2	10.5	11.4
Half Cystine	1.2	0.7	tr	tr
Valine	4.9	5.2	4.7	4.8
Methionine	0.2	0.3	0.4	0.7
Isoleucine	0.9	1.0	1.6	1.0
Leucine	3.8	3.8	3.9	3.7
Tyrosine	0.6	1.0	1.2	0.7
Phenylalanine	2.2	2.1	2.0	1.7

(e = early, l = late stimulation)

NaOH, 1 M $NaBH_4$, 45°C, 30 h) as described by Carlson (1968). Under these conditions most of the carbohydrate is removed from the peptide chain by a beta elimination reaction which splits the O-glycosidic bond between the N-acetylgalactosamine of the oligosaccharide side chain and the hydroxyl group of serine or threonine of the peptide core. In the presence of borohydride the N-acetylgalactosamine involved in the linkage is reduced to N-acetylgalactosaminitol (Fig. 1.3).

The digest was lyophilized, borate was removed as the methyl ester and the sample was chromatographed on a column of Sephadex G-25 to separate the oligosaccharides into fractions of different sizes. Figure 1.4 shows the results of the chromatographic run showing a number of peaks with differing amounts of radioactive sulphate and sialic acid. The peaks of radioactivity and of sialic acid content generally do not coincide. The peaks were pooled as shown, and, after passing the samples through a Dowex 50 column to remove the peptides and glycopeptides, and lyophilizing, the samples were subjected to high voltage electrophoresis at pH 1.9 in a formic acid buffer using a Gilson Electrophorator. The papers were then scanned for radioactivity and stained for carbohydrate by a silver nitrate method (Treyvelyan et al., 1950). At least 85% of the reduced

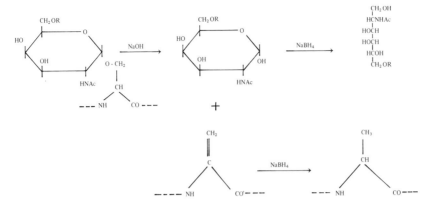

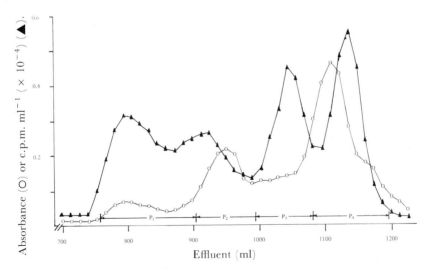

Fig. 1.3   Alkaline borohydride cleavage of O–glycosyl bonds.

oligosaccharides released were acidic and migrated to the positive pole. Figure 1.5 shows the radioactive scan and the carbohydrate-containing zones obtained from the G-25 peaks 2, 3 and 4.

The most abundant fraction containing sulphate (F), and the most abundant one which did not contain sulphate (C) were isolated from peak 4 in milligram quantities by preparative high voltage electrophoresis, and were further purified by descending preparative paper chromatography using butanol, pyridine, water (6:4:3 v/v). The structures of these reduced oligosaccharides were then established from chemical composition, periodate oxidation,

Fig. 1.4   Fractionation of ^{35}S-sulphate labelled oligosaccharides on Sephadex G-25 after alkaline borohydride treatment of canine submaxillary mucin.

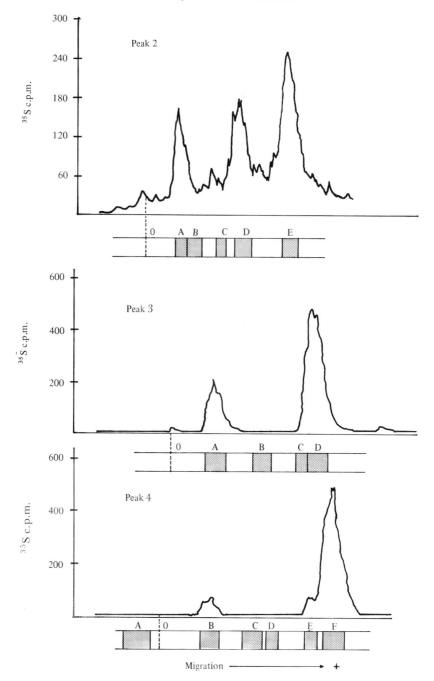

Fig. 1.5 Electrophoretic separation of ^{35}S-labelled and unlabelled oligo-saccharides from peaks P2, P3 and P4 (see Fig. 1.4)

Smith degradation and sequential removal of sugars by specific exoglycosidases.

The structure of oligosaccharides P4C and P4F are shown in Fig. 1.6.

P4C

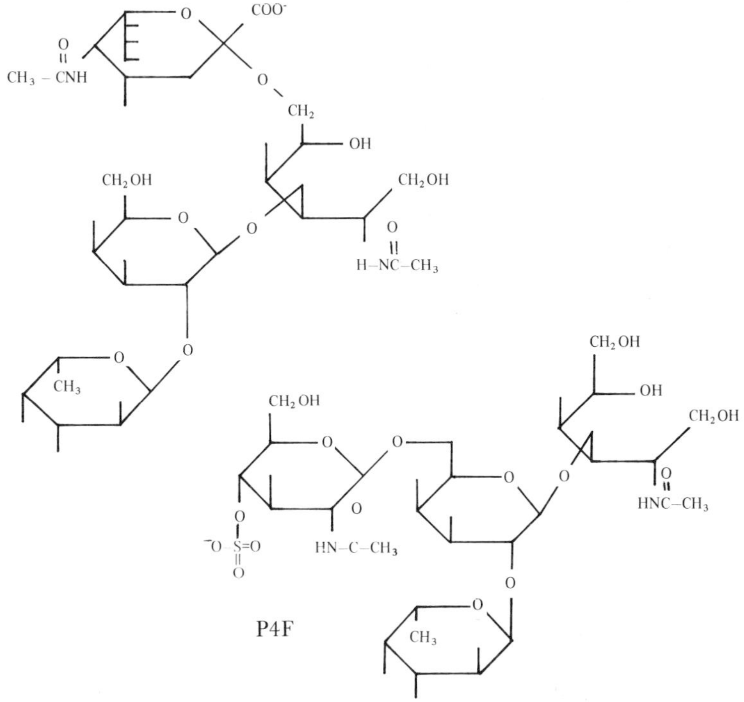

P4F

Fig. 1.6   Oligosaccharide structures P4C and P4F.

The structure of P4C is based on the following information:

a. Sialic acid, fucose, galactose and acetyl galactosaminitol were present in equimolar amounts.

b. Molecular weight by gel filtration was 800. That calculated for the tetrasaccharide is 822.

c. Periodate oxidation revealed 6 vicinal hydroxyl groups per mole—fucose and galactose were destroyed.

d. Smith degradation converted sialic acid to its 7 carbon analogue and converted N-acetylgalactosaminitol to N-acetylthreosaminitol.

e. Neuraminidase removed sialic acid quantitatively and the product upon Smith degradation again yielded threos-aminitol.

f. α-1 → 2 fucosidase released fucose quantitatively.

g. β-galactosidase removed galactose only after fucose had

been released by fucosidase. The product after enzymatic removal of both fucose and galactose, cochromatographed with sialyl-2 → 6-N-acetylgalactosaminitol prepared from ovine submaxillary mucin by treatment with alkaline borohydride.

The structure of P4F is also shown in Fig. 1.6.
This structure is based on following observations.

a. N-Acetylgalactosaminitol, galactose, fucose, N-acetylglucosamine and sulphate were present in equimolecular quantities.
b. Molecular weight by Sephadex G-15 gel filtration was 790, (calculated 815).
c. Periodate oxidation revealed 5 vicinal hydroxyl groups.
d. Fucose and galactose were destroyed by periodate oxidation, but glucosamine was not.
e. Smith degradation converted the N-acetylgalactosaminitol to N-acetylthreosaminitol.
f. β-N-acetyl-glucosaminidase from *Aspergillus niger* (Bahl and Agrawal, 1969) had no influence on the oligosaccharide. However β-N-acetylglucosaminidase isolated from beef liver (Weissmann *et al.*, 1964) split off a glucosamine derivative which was acidic and which was retained on a Dowex 1 Cl⁻ column. Acid hydrolysis of this product yielded free glucosamine and radioactive sulphate.
g. The sulphate was shown to be on the 4 position of acetylglucosamine by reducing with borohydride the sulphated-N-acetylglucosamine released by liver β-N-acetylglucosaminidase. This yielded an N-acetylglucosaminitol sulphate. This was oxidized with periodate, reduced with borohydride and then hydrolysed with 1 N HCl. Analysis with an amino acid analyser showed that all of the glucosaminitol had disappeared and that an equal amount of xylosaminitol had appeared. The sulphate, therefore, had to be in the 4 position.
h. α-1 → 2 fucosidase removed fucose only after treatment with liver β-acetylglucosaminidase.
i. β-galactosidase removed galactose only after both acetylglucosamine and fucose were removed by their respective enzymes.

Preliminary semiquantitative experiments have been carried out to determine the relative amounts of P4C and P4F in the submaxillary mucin isolated from early and late pilocarpine secretions. The P4C:P4F ratio was consistent with the analytical data given previously, i.e. early pilocarpine-stimulated mucin had more of the

sialyated oligosaccharide (P4C) and less of the sulphated oligosaccharide (P4F) than did the late pilocarpine mucin or the electrically-
stimulated mucin.

It was evident from the high voltage electrophoresis experiments of Fig. 1.5 that a number of acidic oligosaccharides were
released from submaxillary mucin by alkaline borohydride. Some
of these were sulphated and some were not. Due to limited amounts
of secreted mucin, the other acidic oligosaccharides from secretions
were not studied. However we have examined the oligosaccharides
released from submaxillary mucins isolated from canine submaxillary
glands. Using much the same methods already outlined, seven
acidic oligosaccharides have been isolated and their structures
determined. These turned to be of two series, one containing sialic
acid and no sulphate (Type A), and the other containing sulphate
and no sialic acid (Type B). The largest Type A oligosaccharide
isolated was the same as the tetrasaccharide P4C isolated from
secreted mucins. The other Type A oligosaccharides were smaller
derivatives (or precursors) of this tetrasaccharide. Their structures
are shown in Fig. 1.7. There are additional sialic acid-containing
oligosaccharides that are larger than the tetrasaccharide, but no
structural work has been done on them.

The Type B series had no sialic acid but contained sulphate.
The structures of four oligosaccharides in this series are shown in
Fig. 1.7.

These included the tetrasaccharide isolated from the secreted
mucin, as well as larger and smaller derivatives or precursors.
A significant observation is the fact that no oligosaccharides containing both sulphate and sialic acid have ever been observed.

It is our hypothesis that the variable ratios of Type A and
Type B oligosaccharides in secreted canine submaxillary mucin
results from changing relative activities of sialyl or acetylglucos-
aminyl transferases or of the concentration of their sugar nucleotide
pools, so that synthesis favours one or the other of the two series of
acidic oligosaccharides. Decision as to whether a sialated or
sulphated oligosaccharide chain is to be formed probably occurs at
the disaccharide stage as suggested in Fig. 1.8.

According to this hypothesis, the addition of sialic acid to the
N-acetylgalactosamine of the disaccharide prevents the addition of
N-acetylglucosamine to the galactose and results in a Type A
oligosaccharide chain. If, on the other hand, a glucosamine is added
to the galactose, sialic acid cannot be added to the N-acetylgalactosamine and a Type B oligosaccharide chain is formed. Both types
appear to be extended by addition of monosaccharides at their
non-reducing termini. Nothing is known about when sulphation of
the N-acetylglucosamine residues occurs. It is interesting that no

REDUCED SIALATED OLIGOSACCHARIDES
FROM CANINE SUBMAXILLARY MUCIN
(TYPE A)

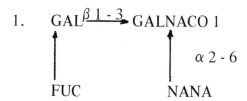

1.

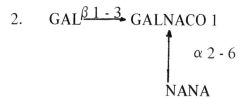

2.

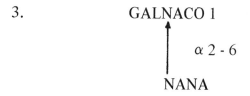

3.

Fig. 1.7

REDUCED SULPHATED OLIGOSACCHARIDES
FROM CANINE SUBMAXILLARY MUCIN
(TYPE B)

1.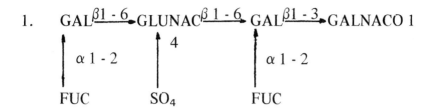

2.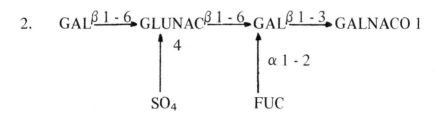

3.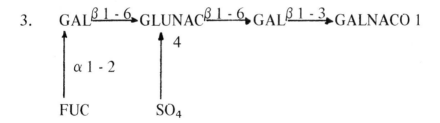

4.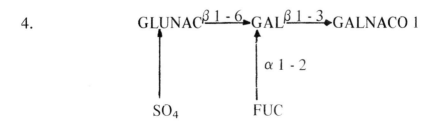

Fig. 1.7   (continued)

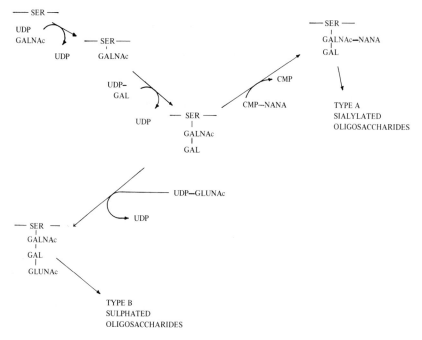

Fig. 1.8 Proposed biosynthetic pathways for sulphated and sialated oligosac-charides.

oligosaccharides containing glucosamine but no sulphate have been found. These would, however, occur in the neutral fraction which we have not yet had the opportunity to investigate.

 If we can consider the system that I have described as a model for membrane glycoprotein biosynthesis it becomes evident that the structure of membrane glycoproteins and thus of membrane-mediated information may depend on the physiological circumstances at the time of membrane biosynthesis.

REFERENCES

Aminoff, D. (1961). *Biochem. J.*, *81*, 384–392.

Antonopoulos, C. A. (1962). *Acta Chem. Scand.*, *16*, 1521–1522.

Bahl, O. P. and Agrawal, K. M. L. (1969). *J. Biol. Chem.*, *244*, 2970–2978.

Barber, A. J. and Jamieson, G. A. (1971). *Biochem. Biophys. Acta*, *252*, 533–539.

Bignardi, C., Aureli, G., Balduini, C. and Castellani, A. A. (1964). *Biochem. Biophys. Res. Comm.*, *17*, 310–312.

Carlson, D. M. (1968). *J. Biol. Chem.*, *243*, 616–626.

Cunningham, L. W. (1971). In *Glycoproteins of Blood Cells and Plasma*, pp. 16–34, ed. Jamieson, G. A. and Greenwalt, T. J.

Dische, Z. (1963). *Ann. N. Y. Acad. Sci.*, *106*, 259–270.

Dische, Z., Pallavicini, C., Kavasaki, H., Smirnow, N., Cizek, L. and Chien, S. (1962). *Arch. Biochem. Biophys.*, *97*, 459–469.

Dubois, M., Gilles, K. A., Hamilton, J. K., Robers, P. A. and Smith, F. (1956). *Anal. Chem.*, *28*, 350–356.

Dunn, J. T. and Spiro, R. G. (1967). *J. Biol. Chem.*, *242*, 5556–5563.

Lehnhardt, W. F. and Winzler, R. J. (1968). *J. Chrom.*, *34*, 471–479.

Lombart, C. and Winzler, R. J. (1972). *Biochem. J.*,

Martinez-Palomo, A. (1970). *Int. Rev. Cytol.*, *29*, 29–76.

Montgomery, R. (1972). In *Glycoproteins. Their Composition, Structure and Function.* 2nd Edition, pp. 518–528, ed. Gottschalk, A.

Rambourg, A. (1972). *Int. Rev. Cytol.*, 31, 57–114.

Roseman, S. (1970). *Chem. and Phys. of Lipids*, 5, 270–297.

Tettamanti, G. and Pigman, W. (1968). *Arch. Biochem. Biophys.*, *124*, 41–50.

Thomas, D. B. and Winzler, R. J. (1969). *J. Biol. Chem.*, *244*, 5943–5946.

Thomas, D. B. and Winzler, R. J. (1971). *Biochem. J.*, *124*, 55–59.

Trevelyan, W. E., Procter, D. P. and Harrison, J. S. (1950). *Nature (London)*, *166*, 444.

Wagh, P. V., Bornstein, I. and Winzler, R. J. (1969). *J. Biol. Chem.*, *244*, 658–665.

Watkins, W. M. (1972). In *Glycoproteins. Their Composition, Structure and Function.* 2nd Edition, pp. 830–891, ed. Gottschalk, A.

Weber, P. and Winzler, R. J. (1969). *Arch. Biochem. Biophys.*, *129*, 534–538.

Weissmann, B., Hadjiioannou, S. and Tornheim, J. (1964). *J. Biol. Chem.*, *239*, 59–63.

Winzler, R. J. (1970). *Intl. Rev. Cytol.*, *29*, 77–125.

Winzler, R. J. (1972. In *Glycoproteins. Their Composition, Structure and Function.* 2nd Edition, pp. 1268–1293, ed., Gottschalk, A.

2 Glycolipids and Cancer

D. R. Critchley
Imperial Cancer Research Fund Laboratories, London

The cell surface is a logical point at which to look for possible causes of the breakdown in cellular interaction characteristic of malignant cells. Attention at this level has been justified by the information from immunological studies, the use of plant lectins, electron microscopy, cell electrophoresis, and chemical analysis. The results of these studies have led to the concept that complex carbohydrates may be involved in the control of cell surface properties important in growth regulation (Kraemer, 1971).

Interest in one class of complex carbohydrate, the glycolipids, began with the observation that homogenates of human tumours were more effective in eliciting antibodies against lipid-type antigens than comparable normal tissue (Rapport, 1969). Such antilipid antibodies reacted more strongly with total lipids of tumour tissue than those from normal tissue. Rapport and co-workers (1958) isolated a pure lipid hapten from human epidermoid carcinoma which reacted with rabbit antisera directed against many different types of human tumour. Analysis showed the lipid hapten to be lactosyl ceramide (cytolipin H). More recently cytolipin R, a ceramide tetrahexoside, was identified as a lipid hapten in rat lymphosarcoma (Rapport, Schneider and Graf, 1967; Rapport and Graf, 1969).

The possible importance of glycolipid antigens in malignancy has been highlighted by the work of Tal et al. (1964, 1965). They originally reported that 90% of the sera of 120 cancer patients agglutinated a suspension of HeLa cells. Only 16% of sera from 51 patients suffering from non-neoplastic chronic disease, and 13% of 237 normal sera showed the effect. All of 12 sera from pregnant women showed agglutination. Several other tumour cell suspensions were agglutinated although normal liver or kidney cells were unaffected. The factor was absorbed from serum by lactosyl ceramide but not by other glycolipids having similar structure, and the agglutination was inhibited by lactose. The use of this system in the diagnosis of cancer has been discussed (Tal and Halperin, 1970).

Aberrations in metabolism of more complex glycolipids related to blood group activities have been found in human adeno-carcinomas by Hakomori *et al.* (1967). A tumour glycolipid was isolated which lacked blood group A or B activity regardless of the blood type of the tissue donor, but had a weak H and consider-able Le^a activity. Its structure was found to be β-gal-(GlcNAc)-gal-glu-cer with a fucosyl residue attached at the penultimate sugar. Subsequently Le^b active glycolipid and a glycolipid with unknown specificity have been found in considerable quantities in tumour tissue, although no A and B activities have been detected (Yang and Hakamori, 1971; Hakamori, 1970a). Changes in glyco-lipid pattern have also been found in human brain tumours which were characterized by elevated levels of hematoside and disialo-hematoside in contrast to the predominance of higher ganglioside in normal brain (Siefert and Uhlenbruk, 1965).

Although these limited number of studies suggested the involve-ment of glycolipid changes in malignancy, it should be noted that at least one situation has been found in which human cancer tissue, a hepatoma, had apparently normal glycolipid metabolism (Kaw-anami and Tsuji, 1968).

FACTORS AFFECTING GLYCOLIPID COMPOSITION *IN VIVO*

Glycolipids are thought to be relatively tissue specific compared to, say, phospholipid, although they are also affected by species, genetic strain, sex, age and pathological state of the animal.

Compared to brain, extraneural organs contain small amounts of glycolipid but there are distinct differences in the proportion of neutral glycolipids, sulphatides and gangliosides between organs (Martensson, 1969). There are also variations from tissue to tissue within each group, e.g. the neutral glycolipid of brain is predomi-nantly monohexosyl ceramide, in spleen it is dihexosyl, and kidney tetrahexosyl ceramide (Martensson, 1969). Species variation is exemplified by the decrease in concentration of gangliosides in the brains of fish, compared to birds and mammals (Svennerholm,

1970). Strain differences in total glycolipid of the kidney of mice have been reported by Coles *et al.* (1970). They also found sex of the animal to be a key factor. Little diglycosyl ceramide was accumulated in kidneys of female mice compared to males of the same strains, although administration of testosterone to females induced its synthesis (Gray, 1971). Changes in the ratios of monohexosyl ceramide and its sulphatide occur in mouse kidney with age (Coles *et al.*, 1970), and changes in brain glycolipids have been shown to parallel the onset of myelination in the foetus (Svennerholm (1964, 1970)).

Glycolipid metabolism is also known to be modified in certain pathological conditions. The hereditary diseases collectively termed sphingolipidoses are characterized by an increased concentration of various glycolipids of certain tissues (Brady, 1970) and the kidney glycolipids of certain strains of mice are also changed in the presence of strain specific ascites tumours (Adams and Gray, 1967). The sex of the mouse and type of tumour both apparently determined the modification (Coles *et al.*, 1970).

The sensitive metabolic control and their antigenic properties apparently expressed at the cell surface make glycolipids of great interest with respect to a study of the nature of the change of the surface membrane in malignancy. Comparison of normal and tumour tissue suffers from the disadvantage that tissues are heterogenous with respect to cell types. The ability to transform relatively stable cell lines in culture with tumour viruses has provided a new approach to cancer biochemistry. Tumour viruses have been shown to produce a number of genetically stable changes in the cells they transform. The main interest in these changes is that one or more may be responsible for a cell *in vivo* becoming tumorigenic. The properties deserving special attention are those occurring regularly in independently transformed clones, especially if they are common to cells transformed by, say, a small DNA tumour virus and a large RNA virus. Such differences are more likely to be relevant to the common ability these viruses have to induce tumours than other properties unique to transformation by individual viruses.

GLYCOLIPIDS OF CULTURED CELLS

A comprehensive review of glycosphingolipid metabolism is beyond the scope of this review and biosynthetic pathways are included only to aid appreciation of some of the sequences demonstrated in various tissues which may possibly operate in cultured cells (Fig. 2.1). The subject has been the topic of several recent reviews (Svennerholm, 1970; Morell and Braun, 1972).

Partial characterization of the glycolipids of a variety of hamster (Hakomori and Murakami, 1968; Sakiyama, Gross and

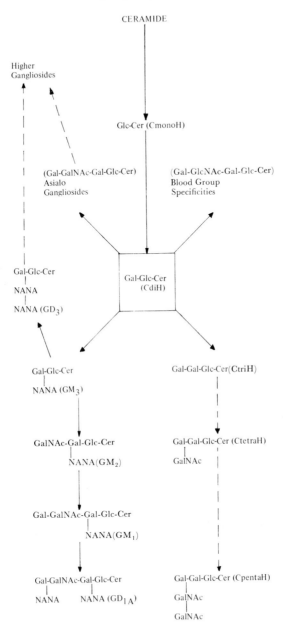

Fig. 2.1 Possible biosynthetic inter-relationships of the glycolipids commonly found in cultured cells emphasizing the key positions of ceramide dihexoside. The abbreviations CmonoH to CpentaH are used to denote neutral glycolipids with 1–5 sugar residues. For ganglioside nomenclature see Svennerholm, 1970. Unbroken lines, synthetic pathway demonstrated; dashed lines, pathway assumed.

Robbins, 1972), mouse (Hakomori, Teather and Andrews, 1968; Weinstein *et al.*, 1970), rat (Dawson, Matalon and Dorfman, 1972), bovine, monkey (Klenk and Choppin, 1970), chicken (Hakomori, Saito and Vogt, 1971b; Warren, Critchley and Macpherson, 1972), and human cells (Hakomori, 1970b; Dawson, Matalon and Dorfman, 1972), in culture have been reported (Table 2.1). Hematoside (GM_3) is a major component of most of the cells, but analysis suggests that those cells which accumulate significant amounts of neutral glycolipid do not contain the higher gangliosides.

That the glycolipids are synthesized by the cells was examined in L 929, and mouse embryo secondary cultures by Yogeeswaran *et al.* (1970), in view of the fact that cultured cells derive much of their lipid from serum (Howard and Kritchevsky, 1969; Rothblat and Kritchevsky, 1966). They calculated that there was almost enough ganglioside in the culture medium to account for all the cell ganglioside. The pattern of serum ganglioside was however, different from the cell. In addition, studies with radioactive isotopes show that glucose (Hakomori, 1970b), galactose (D. Critchley, unpublished), glucosamine (Hakomori, 1970b; Yogeeswaran, Sheinin, Wherrett and Murray, 1972; Yogeeswaran, Wherrett, Chatterjee and Murray, 1970; Hakomori, Saito and Vogt, 1971b), N-acetylmannosamine (Brady and Mora, 1970) and serine (Hakomori, 1970b), are all incorporated into cellular glycolipid. In our studies (Critchley and Macpherson, 1973) with $[1\text{-}^{14}C]$palmitate, 50% of the label in the glycolipid molecule was incorporated into the sphingosine base, and 50% into the fatty acid, indicating synthesis of whole molecules of glycolipid by the cell. However, as shown for erythrocytes (Marcus and Cass, 1969), cultured cells can take up glycolipid from the medium. In experiments in which NIL 2 cells were grown in the presence of 50 μg/ml CtriH, 30–50% uptake was observed over a 72 h period in which the cells had multiplied from 0.5–5 $\times$ 10^6 cells per 9 cm dish (D. Critchley, unpublished). However, most of the CtriH had been metabolized to non-glycolipid products. The fact that the glycolipid composition of mouse cell types is distinct from that of hamster suggests that the pattern is indeed genetically determined. The recent demonstration that some of the glycolipid glycosyl transferases are present in cultured cells (Cumar, Brady, Kolodny, McFarland and Mora, 1970; Fishman, McFarland, Mora and Brady, 1972; Kijimoto and Hakomori, 1971; Den, Shultz, Basu and Roseman, 1971), confirms that the cell at least has the potential to synthesize its own glycolipids. The complete pathway for synthesis of GD_{1A} from hematoside has been demonstrated in 3T3 cells (Fishman *et al.*, 1972). The possible influence of serum glycolipids on endogenous synthesis should however be remembered.

Table 2.1
GLYCOLIPID COMPOSITION OF CULTURED CELLS

Cell density	Cell Type	CmonoH	CdiH	CtriH	CtetraH	CpentaH	GM₃	GD₃	GM₂	GM₁	GD₁ₐ	Total	Ref.
	Hamster												
C	BHK-21	40	50	300	—	—	520	50	—	<30	—	990	1
C	BHK-21	90	90	—	—	—	560	—	—	—	—	740	1
C	BHK-21	42	23	36	—	—	325	—	—	—	—	426	2
C	NIL2	—	12	40	37	80	140	—	—	—	—	241	3
C	NIL2	72	ND	ND	196	407	120	—	—	—	—	795	4
	Mouse												
ND	3T3	15	<5	—	—	—	177	—	—	—	—	197	5
C	3T3	17	5	—	—	—	132	—	—	—	—	164	2
G	3T3	ND	ND	—	—	—	465	—	ND	113	575	1153	6
G	Balb 3T3	ND	ND	—	—	—	122	—	—	48	154	324	6
G	AL/N	ND	ND	—	—	—	71	—	—	245	345	661	6
C	Neuroblastoma	27	31	—	91	—	6	—	115	28	125	423	2
	Others												
C	Rat Glial	39	9	—	25	—	220	23	—	—	—	316	2
ND	Bovine kidney	160	—	—	510	—	9	—	—	—	—	679	7
C	Chick embryo fibroblast	10	—	—	—	—	575	100	—	150	75	925	8
C	Human diploid fibroblast	100	—	—	—	—	550	380	—	250	—	1280	1
C	Skin fibroblast	32	18	73	42	—	134	30	—	—	—	329	2

Data expressed as μg glycolipid/100 mg cell protein. Cell density refers to either growing (G) or confluent (C) cells; ND no data given.
Refs for Table 2.1.

1. Hakomori (1970*b*). 2. Dawson, Matalon and Dorfman (1972). 3. Kijimoto and Hakomori, (1972). 4. Sakiyama, Gross and Robbins, (1972). 5. Hakomori, Teather and Andrews (1968). 6. Brady and Mora (1970). 7. Klenk and Choppin (1970). 8. Hakomori, Sata and Vogt, (1971).

Whether the glycolipids that accumulate in the cells are precursors of the more complex glycolipids has received little consideration. It is possible for example, that the CtriH which accumulates in NIL2 cells is not itself the precursor of the tetra or penta-hexosides. The UDP N-acetylgalactosaminyl transferase may have specificities for sphingosine base or fatty acid and only utilize a specific part of the trihexoside pool.

In this context Weinstein *et al.* (1970), found differences in the fatty acid composition of various L cell glycolipids. Ceramide dihexoside and hematoside of whole cells contained predominantly short chain saturated fatty acids. Monosialoganglioside contained slightly more unsaturated fatty acid but disialoganglioside contained 34% unsaturated and 45% long chain fatty acids. The fatty acid content of sphingomyelin was also totally different from that of glycolipid suggesting that they are not synthesized from the same precursor pool of ceramide.

Similar situations with respect to glycolipid fatty acids have been reported for 3T3 cells (Yogeeswaren *et al.*, 1970) and rat liver (Siddiqui and Hakomori, 1970). However, the possibility that acyl chain rearrangement takes place via a ceramidase cannot be discounted.

In contrast the similarity in fatty acid content of BHK21 hamster glycolipids (predominantly $C_{16:0}$, $C_{18:0}$ and $C_{18:1}$) was suggested to support the idea of precursor product relationships (Dawson, Matalon and Dorfman, 1972).

GLYCOLIPIDS OF CELLS BEFORE AND AFTER TRANSFORMATION WITH TUMOUR VIRUSES

The first experiments comparing the glycolipids of normal and virus transformed cells were those of Hakomori and Murakami, utilizing BHK21 (C13) (Hakomori and Murakami, 1968). In the original weakly tumorigenic contact inhibited clone, sialyllactosyl ceramide (hematoside) was the major glycolipid with small amounts of lactosyl ceramide (Table 2.2). Transformation of the line by polyoma virus (Py) produced a non-contact inhibited highly tumorigenic line in which the level of hematoside was reduced (4 times) and the lactosyl ceramide increased (10 times). The overall levels of glycolipid per protein base were reduced by half. A spontaneously transformed line with intermediate patterns of contact inhibition and tumorigenicity had a glycolipid pattern between that of the normal and transformed lines.

This exciting result was in agreement with the results on blood group substances (Hakomori, 1970a) which suggested that the carbohydrate chain in tumour cells is often less complete than those in non-tumorigenic cells. Interestingly, the degree of agglutination of the cell types by wheat germ agglutinin parallelled their malignant properties and the small amount of higher ganglioside present in transformed cells inhibited the interaction. Normal BHK21 higher gangliosides were ineffective but removal of the terminal sugar by Smith degradation produced an effective inhibitor. This again suggested that the surface properties of transformed cells may be altered due to incompletion of the carbohydrate chains.

Further studies by Hakomori *et al.* (1968) have confirmed these results and extended them to BHK transformed by Rous sarcoma virus (Table 2.2). The overall reduction in total glycolipid and

Table 2.2
GLYCOLIPIDS OF NORMAL AND TRANSFORMED BHK21
AND 3T3 CELL LINES

Cell type	GM_3	CdiH	CmonoH	Total
BHK-21.C13 (clone 1)	475*	13	N.D.	488
Spontaneous transformed cell (clone 2)	310	117	N.D.	427
Polyoma Virus transformed cell (clone 3)	105	125	N.D.	225
BHK-21.C13	300	50	—	350
Transformed by Schmidt Rupin type Rous virus	60	65	—	125
Transformed by Bryan type Rous	75	80	—	155
3T3	177	<5	15	192
Transformed by polyoma virus	82	<5	20	102
Transformed by SV40	63	<5	—	63
Doubly transformed	27	<5	30	57

* μg glycolipid/100 mg cell protein.
Abstracted from S. Hakomori and W. T. Murakami (1968) and S. Hakomori *et al.* (1968).

decreased hematoside concentration were still marked although the increase in lactosyl ceramide appeared more variable. The variable aspect of the change in BHK Py was noted in a subsequent study in which hematoside was reported to decrease but the lactosyl ceramide was unchanged or rather elevated (Hakomori, Kijimoto and Siddiqui, 1971a). Similar results were reported for 3T3 mouse embryo fibroblasts transformed by polyoma and SV40 viruses and a human heteroploid line transformed by SV40; again the major change was decreased hematoside (Hakomori, Teather and Andrews, 1968).

A more detailed examination of the effects of transformation on the glycolipids of mouse cell lines was reported by Mora *et al.*

(1969, Brady and Mora, 1970) using 3T3 cells and AL/N mouse embryo epithelial cells. In contrast to the observations of Hakomori *et al.* (1968), they detected large quantities of mono and disialogangliosides in normal cells. These were markedly reduced in SV40 and polyoma transformed cell lines, although hematoside remained unaffected. In essence therefore, the results were consistent with the idea that there is incomplete synthesis of carbohydrate chain in transformed cells. The data have been confirmed in the Balb 3T3/SV40 system (Brady and Mora, 1970; Dijong *et al.*, 1971). In addition no modification of the ganglioside pattern was observed in the spontaneously transformed highly tumorigenic TAL/N or Balb 3T 12/1 lines. The modification of ganglioside pattern on viral infection has been rationalized by the finding that the activity of the enzyme catalyzing the addition of N-acetylgalactosamine to hematoside is markedly reduced in transformed cells (Cumar, Brady, Kolodny, McFarland and Mora, 1970; Fishman, McFarland, Mora and Brady, 1972; Mora, Cumar and Brady, 1971). The possibility that catabolic processes were increased in transformed cells was partially excluded (Cumar *et al.*, 1970). That the effect is dependent on a functioning integrated viral genome was shown by the near normal synthetic capacities of flat revertants of polyoma and SV40 transformed 3T3, and 3T3 and AL/N cells lytically infected with polyoma virus (Mora, Cumar and Brady, 1971).

A similar block in the capacity of the BHK Py to synthesize hematoside from lactosyl ceramide has been shown (Den, Schultz, Basu and Roseman, 1971). It is interesting to note that the enzyme affected is apparently cell- rather than virus-specific. Thus the mechanism is not a specific modification inserted at a particular point in the cell's biochemical machinery like that imposed on Salmonella O-antigen polymerase by a series of temperate bacteriophages (Losick and Robbins, 1969).

GLYCOLIPID PATTERN IN MORRIS HEPATOMAS

Siddiqui and Hakomori (1970), have extended their observations to the Morris hepatoma system in rats. A comparison of normal rat liver, rapidly dividing neonatal liver and three hepatomas gave a complete spectrum of rates of cell division, one of the hepatomas growing at a slower rate than neonatal liver. The normal adult liver contained GM_3, GM_1, GD_{1A} and GT_1. Interestingly, the neonatal glycolipids contained a greater preponderance of the trisialoganglioside than adult liver so that rapid cell division per se is not the limiting factor in the synthesis of higher gangliosides (Table 2.3). In contrast the hepatomas all consistently lacked significant amounts of trisialoganglioside although there was a

Table 2.3
GLYCOLIPID PATTERNS OF NORMAL LIVER AND
MORRIS HEPATOMAS

	Normal Adult Rat Liver	Baby Rat Liver	Tumour 5123 (rapid)	Tumour 7800 (slow)
Ceramide	0.021*	N.D.	0.43	0.41
Glucosyl Cer.	0.33	N.D.	0.93	0.83
GM_3	0.13	0.21	0.64	0.57
GM_1	0.11	0.04	2.57	1.87
GD_{1A}	0.09	0.15	3.60	1.73
GT	0.01	0.09	0	0

* m μmoles/mg protein.
Abstracted from B. Siddiqui and S. Hakomori (1970).

marked accumulation of GD_{1A}. The tumour lines also contained more total ganglioside/mg protein and elevated levels of ceramide and lactosyl ceramide. Thus although the tumours contain insignificant quantities of GT_1, total ganglioside content is much higher than in normal adult or neonatal liver. The data has been suggested to support the thesis that a decrease in levels of the more complex glycolipids accompanies chemically induced malignancy.

Essentially similar observations were made by Cheema *et al.* (1970), in a qualitative study. Comparison of the ganglioside patterns of an hepatic epithelial line and an hepatoma cell line has also been made *in vitro* (Brady, Borek and Bradley, 1969). The normal hepatocytes contained GM_3, GM_1 and GD_{1A}. The level of GD_{1A} was decreased 4–7 times in the hepatocyte line although as previously observed by Siddiqui and Hakomori (1970) in their *in vivo* studies, the total ganglioside content/mg protein increased. This latter observation is in marked contrast to those on virus transformed cells where the reduction in complexity of the carbohydrate chains of glycolipid is accompanied by an overall decrease in glycolipid concentration.

CELL DENSITY DEPENDENT GLYCOLIPID SYNTHESIS
The possibility that glycolipid metabolism is in some way linked to growth control mechanisms received further support from the observation that low passage BHK, with low saturation density and tumorigenicity, contained a ceramide trihexoside, the level of which was 2–3 times greater in dense, compared to sparse cultures (Hakomori, 1970b). BHK of higher passage, which reached higher saturation densities and were more tumorigenic, did not contain the component. In addition, transformation of the low

passage BHK lead to complete loss of the trihexoside. The quantitative results were confirmed by studying the incorporation of radioactive serine and glucose.

In a similar study of the NIL2 Syrian Hamster embryo line, Robbins and Macpherson (1971a, b), noted the increased incorporation of $[1\text{-}^{14}C]$palmitate into a ceramide tri-, tetra-, and an unknown (now thought to be pentahexoside) in dense, compared to sparse, cultures. These density dependent glycolipids were not present in transformed cells, which contained only those glycolipids not showing the phenomenon, i.e. CmonoH, CdiH, hematoside (Table 2.4).

Table 2.4

INCORPORATION OF $[1\text{-}^{14}C]$ PALMITATE INTO LIPIDS OF NIL2 AND NIL2/HSV CELLS IN SPARSE AND DENSE CULTURE

Cells were labelled for 48 h with 10 μc $(1\text{-}^{14}C)$ palmitate; PC + SM phosphatidyl choline plus sphingomyelin, PE phosphatidyl ethanolamine, PI phosphatidyl inositol. HSV, Hamster Sarcoma Virus.

Lipid	NIL2		NIL2/HSV	
	Sparse	Dense	Sparse	Dense
PC + SM	680*	615	742	690
PE	143	198	113	142
PI	48.6	53.0	15.5	56.5
CmonoH	15.6	5.5	26.5	4.0
CdiH	7.9	6.7	17.6	8.6
CtriH	2.5	18.2	0	0
CtetraH	2.0	7.6	0	0
CpentaH	5.4	11.3	0	0
GM$_3$	45.0	30.2	48.0	34.0
$\dfrac{GM_3}{CtriH}$	18.0	1.6		

* Results represent cpm $\times$ 10^3/cpm in phospholipid plus glycolipid.
D. Critchley and I. Macpherson (1973).

A density dependent disialohematoside and monosialoganglioside were also found in a human fibroblast line by Hakomori (1970b), although similar changes have been suggested to be absent in 3T3 cells (Fishman et al., 1972).

The increased levels of certain glycolipids in dense compared to sparse culture, are partially explained by the observation that there is a 2–3 fold increase in the activity of a UDP gal:lactosylceramide α-galactosyl transferase in dense cultures (Kijomoto and Hakomori,

1971) (Table 2.5). The enzyme was barely detectable in the transformed cells in contrast to the β-galactosyl transferase which catalyzes conversion of ceramide mono to dihexoside. This enzyme showed little change in activity with cell density or after transformation. In addition the α-galactosidase was found to be somewhat elevated in transformed cells although no density dependence was demonstrated.

Table 2.5

ACTIVITIES OF UDP-gal: GLYCOLIPID α- AND β-GALACTOSYL-TRANSFERASES OF A SUBCELLULAR FRACTION FROM NORMAL AND TRANSFORMED BHK21 AND NIL CELLS

	Cell population densities	Glycolipid synthesized in $\mu\mu$moles/mg protein/hour (complete system)	
		CtriH synthesis	CdiH synthesis
BHK	sparse ($\leqq 5 \times 10^4/cm^2$)	109 (4)	122
	confluent ($> 10^5/cm^2$)	385 (5)	167
*PY-BHK	Low ($\leqq 10^5/cm^2$)	55	
	High ($> 10^5/cm^2$)	48 (4)	121
NIL-2E	sparse ($\leqq 5 \times 10^4/cm^2$)	385 (2)	247 (2)
	confluent ($> 10^5/cm^2$)	1052 (2)	257 (2)
*PY-NIL	Low ($\leqq 10^5/cm^2$)	26	400 (2)
	High ($> 10^5/cm^2$)	38 (2)	206

* PY: polyoma transformed.

Following brief sonication, the supernatant from centrifugation at 12,000 g × 15 min was sedimented at 105,000 g × 1 h. This fraction was used in the assay. Values for the incomplete assay system were:

10–30 $\mu\mu$moles/mg h minus CdiH or CmonoH;

10–20 $\mu\mu$moles/mg/h for minus Mn^{++}.

Abstracted from S. Kijimoto and S. Hakomori (1971).

A study of the kinetics of synthesis of the compounds in NIL2 cells (Critchley and Macpherson, unpublished), is shown in Fig. 2.2. An increase in cell number from 2×10^6 per 9 cm dish, in which there is little cell contact, to 5×10^6, resulted in a sharp increase in incorporation of $[1\text{-}^{14}C]$palmitate into CtriH compared to GM_3, even though the cells are still capable of further division.

In the reciprocal experiment where cells were plated to low density from dense culture by trypsinization, decreased incorporation of label into the density dependent glycolipids occurred before the first cell division, although it took between 10 and 18 h for the change to occur (Fig.2.3). Cells which were trypsinized and reseated

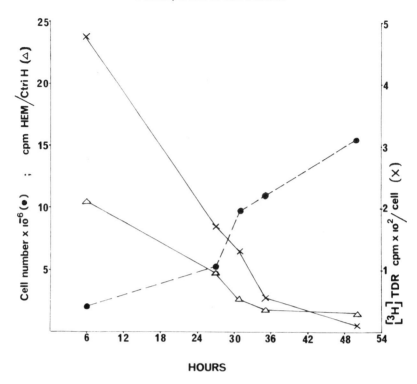

Fig. 2.2 Synthesis of glycolipids in NIL2 cells growing from sparse to dense culture. Cells were plated at $2 \times 10^5/9$ cm dish. At 2×10^6 cells (still little cell contact) 4–8 dishes were pulsed with 10 μc of $[1\text{-}^{14}C]$ palmitate (6 h), 2 with 0.1 μc/ml. $[^3H]$ thymidine (6 h), and 2 dishes were taken for cell number estimation. Similar points were taken as cell densities increased. The data relates the incorporation of $[1\text{-}^{14}C]$ palmitate into hematoside (GM_3)/CtriH ($\triangle$–$\triangle$), cell number ($\bullet$–$\bullet$), and $[^3H]$ thymidine (TDR) incorporation (X–X).

at high cell density showed a small but significant decrease in incorporation into CtriH compared to GM_3, which returned to control values after 34 h. Experiments on turnover did not suggest that the pre-existing pool of density dependent compounds was rapidly degraded before the first cell division.

That the increased synthesis is not just a property of non-dividing cells was also tested by blocking cell division under conditions where there was little or no cell contact (Table 2.6). Cells in two of the situations tested failed to synthesize the compounds, although cells blocked by serum depletion incorporated significantly more $[1\text{-}^{14}C]$palmitate into CtriH than logarithmically growing sparse cultures. This discrepancy may reflect variations in synthesis

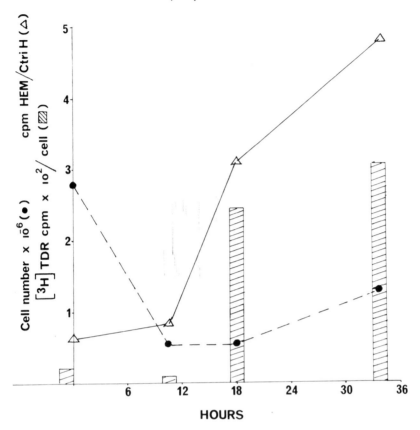

Fig. 2.3 Glycolipid synthesis in NIL2 cells seeded from dense to sparse culture. Dense cell cultures were checked for (^{3}H) thymidine incorporation (0.1 μc/ml, 1 h) and stability of cell counts, over a 2-day period. At quiescence some cells were pulsed with 10 μc [1-^{14}C] palmitate (6 h) and others trypsinized (0.25% trypsin in tris/saline) and plated at 5×10^5 cells/9 cm dish. Dense cultures were trypsinized and directly reseated as a control. Incorporation of [1-^{14}C] palmitate into hematoside (GM$_3$)/CtriH ($\triangle$–$\triangle$) was related to cell number ($\bullet$–$\bullet$) and (^{3}H) thymidine (TDR) incorporation (▨) at points before and after the first cell division.

in different parts of the cell cycle. Non-dividing transformed cells did not regain the ability to synthesize CtriH.

The possibility that transformation of the NIL2 cells was selecting for a section of the population which lacked the density dependent components was considered unlikely. Extensive cloning of our NIL2 cells has failed to produce a line lacking the density dependent glycolipids, although some variation in the proportion of the [1-^{14}C]palmitate label in these compounds is noticeable from clone to clone (Table 2.7). More dramatic clonal variation in

Table 2.6

DISTRIBUTION OF (1-^{14}C) PALMITATE IN THE GLYCOLIPIDS OF GROWTH INHIBITED NIL2 CELLS

Cells from strictly low density cultures were plated in medium (a) containing only 0.25% serum; (b) lacking glutamine; (c) containing 2 mM thymidine. The onset of the effect of the inhibitor was tested by cell counts or [^{3}H] thymidine incorporation. The quiescent sparse cultures were cultured for a further 48 h after the addition of 5–10 μc [1-^{14}C] palmitate.

Lipid	Cells in sparse culture	High density	Cells inhibited by: Low serum	No glutamine	Excess thymidine
CmonoH	16*	5	24	23	19
CdiH	25	6	6	13	9
CtriH	6	34	20	8	10
CtetraH	2	17	8	3	6
CpentaH	6	10	8	11	17
GM$_3$	44	28	33	41	37
$\dfrac{GM_3}{CtriH}$	7.3	0.8	1.6	5.0	3.7

* Results are expressed as a percentage of total incorporation into glycolipid. (D. Critchley and I. Macpherson 1973).

Table 2.7

INCORPORATION OF (1-^{14}C) PALMITATE IN GLYCOLIPIDS OF NIL2 CELLS AND VARIANTS IN DENSE CULTURE

Cell lines were labelled with 10 μc (1-^{14}C) palmitate in dense culture for 48 h. Results are expressed as a percentage of incorporation into total glycolipid. HSV, hamster sarcoma virus.

Cell type	CtriH	CtetraH	CpentaH
NIL2 clones no. 1–6	23.5	26.5	21.6
NIL2 clones no. 7–12	15.0	12.5	30.5
NIL2/HSV clones no. 1–6	0.9	1.6	7.0
7–11	0	0	3.0
NIL2/HSV FUDR clones no. 1–6	0	0	3.0
Tumour lines no. 1–10	0	0	20.5
Tumour line no. 11	22.7	17.5	27.2
Tumour line no. 11 (Passaged in hamsters)	19.5	5.4	32.5
Tumour line no. 11 (Transformed by HSV)	1.2	1.5	13.0

(D. R. Critchley and I. Macpherson 1973).

labelling pattern was observed by Sakiyama *et al.* (1972), who isolated one clone of NIL2 cells which failed to synthesize all three compounds. In this clone hematoside was density dependent and significantly the effect was abolished by transformation. They also found no correlation between the presence of the compounds and the saturation density of a line, and in fact the line with all three neutral glycolipids had the highest saturation density.

Extensive cloning of our transformed cells has failed to produce a transformed derivative with the ability to synthesize the compounds (Table 2.7). Flat 'revertants' obtained by FUDR selection were still wild type with respect to glycolipids. This is in contrast to the results in the mouse cell system where Brady and Mora (1970) reported a recovery of normal gangliosides in 3T3 Py and SV40 revertants (Pollack *et al.*, 1968). It is interesting in this context that Yogeeswaran *et al.* (1972) have recently reported two clones of 3T3 SV40 which were phenotypically transformed, still possessed a strong T-antigen, but had a normal ganglioside pattern.

The problem of clonal variation in the cultured cell system may well explain the difference between the results of Hakomori *et al.* (1968) and Brady and Mora (1970) on the ganglioside of 3T3 cells. In addition, we have been unable to confirm the presence of significant amounts of ceramide trihexoside in most sublines of BHK cells and find the differences on transformation with polyoma virus variable, although SV40 transformed BHK consistently showed elevated levels of lactosyl ceramide and reduced hematoside (Wiblin and Macpherson, 1972).

GLYCOLIPIDS OF CHICK EMBRYO FIBROBLASTS— TRANSFORMATION WITH ROUS SARCOMA VIRUS

Use of primary and secondary cultures of chick embryo fibroblast to study the effects of transformation has several advantages. Better than 90% transformation of the culture occurs 3 days post infection with Rous sarcoma virus, due to a continuous process of infection, release of virus, and infection of neighbouring cells. The rapidity of transformation allows comparison of normal and transformed cells before prolonged culture has induced modifications unrelated to virus induced transformation. The time course of transformation changes can also be studied with the added advantage that well characterized temperature sensitive mutants of Rous sarcoma virus are available (Martin, 1970). Studies on transport and agglutination in this system have already been reported (Martin, Venuta, Weber and Rubin, 1971; Eckhart *et al.*, 1971).

In a study on glycolipids of confluent primary cultures, Hakomori, Saito and Vogt (1971b) showed a 2–3 fold reduction in concentration of the major component hematoside in transformed cells (Table 2.8). Larger reductions (10 times) were found in the disialohematoside. The modifications were accompanied by 2–3 fold

Table 2.8

THE EFFECT OF TRANSFORMATION ON THE GLYCOLIPIDS OF CHICK EMBRYO FIBROBLASTS

Glycolipid	NORMAL C.E.F.		VIRUS INFECTED CELLS		
	sparse	dense	dense PR-RSV-C†	dense f. RSV	dense RAV-1
Ceramide	15*	15	50	25	13
CmonoH	10	10	35	25	10
GM$_3$	300	575	170	250	350
GD$_3$	75	100	10	10	125
monosialoganglioside	100	150	15	15	75
disialoganglioside	60	75	50	50	50
TOTAL	560	925	320	375	613

* μg glycolipid/100 mg cell protein.
† Types of Rous sarcoma virus
Abstracted from S. Hakomori, T. Saito and P. K. Vogt (1971b).

increases of ceramide and its monohexoside. Comparison of chick embryo fibroblasts infected with RAV-1, a non-transforming virus, also showed some reduction in the hematoside and monosialoganglioside levels although the other components were not affected. Time course studies suggested that disialohematoside was first affected and levels of hematoside, ceramide and the monohexoside were not significantly modified until transformation was complete.

Of interest in relation to studies on density dependent glycolipids, was an indication that those gangliosides affected by transformation were density dependent in the normal cell.

We have also studied (Warren, Critchley and Macpherson, 1972) this culture system in our laboratory with the hope of utilizing the temperature sensitive Rous sarcoma virus of Martin (1970). However, employing the same method of $[1-^{14}C]$palmitate labelling as has been successfully used to elucidate changes in other cell glycolipids after transformation (Robbins and Macpherson, 1971a, b; Critchley and Macpherson (1973); Sakiyama, Gross and Robbins, 1972), we were unable to detect significant reproducible differences in incorporation into hematoside or the ceramide monohexoside after transformation.

Possible explanations for the discrepancy lie in: (1) our use of $(1-{}^{14}C)$palmitate to reflect glycolipid pattern which assumes that the specific activities of the glycolipids is uniform after a 48 h labelling period (Critchley and Macpherson, 1973); (b) our use of secondary not primary cultures; (c) our comparison of dividing cultures rather than dense cultures in which the normal cells are probably slowly dividing compared with the transformed cells. The demonstration that cell surface glycoproteins were modified by transformation (Warren, Critchley and Macpherson, 1972), but glycolipids were apparently unaffected, may be significant to a study of the interrelationships of synthesis of the two types of complex carbohydrate molecules.

SUBCELLULAR DISTRIBUTION OF GLYCOLIPIDS

That the glycolipid changes in cancer cells are relevant to cell surface properties has until recently largely rested on immunological data which shows that cellular glycolipids can function as antigens (Rapport *et al.*, 1958; Tal, 1965). More direct evidence for a surface location came from the finding that glycolipids were important components of the erythrocyte (Sweeley and Dawson, 1969) and synaptosomal (Weigandt, 1967) membrane. This was extended to rat liver plasma membrane by Dod and Gray (1968a, b). The ratio of mμmoles of glycolipid/μmoles of phospholipid phosphorus was 1.3 for the total liver, 7.1 for the plasma membrane, 0.79 for endo-plasmic reticulum and 0 for mitochondria. The small amount of glycolipid in the endoplasmic reticulum was thought to be due to contamination by plasma membrane.

More recent studies have tended to confirm the idea of a concentration of cellular glycolipid at the surface membrane. Klenk and Choppin (1970) in a study on bovine kidney (MDBK) cells found an 8 fold concentration of CmonoH, and 4 fold concen-tration of CtriH and ganglioside in plasma membrane when compared to whole cells on a protein basis. A similar concentration of ganglioside was found in BHK21-F cells although no comparison in enrichment of other lipids was made. In a comparison of enzymat-ically and antigenically characterized subcellular fractions of BHK21-C13 Renkonen and coworkers (1970, 1972) found a 10 fold enrichment of ganglioside in plasma membrane compared with whole cells on a protein basis. In addition the concentration of ganglioside in plasma membrane was 5–7 times greater than in endoplasmic reticulum. Molar ratios of total glycolipid to phospho-lipid were found to be 0.045 for the cell, 0.076 for the plasma membrane and 0.023 for endoplasmic reticulum, Table 2.9. No data were given for mitochondria or nuclei. If it is assumed that

Table 2.9
LIPID TO PROTEIN RATIOS OF THE PRINCIPAL LIPID FRACTIONS IN BHK21-C13 CELLS, PLASMA MEMBRANES AND ENDOPLASMIC RETICULUM

Lipids	Cell	Plasma Membrane	Endoplasmic Reticulum
Total lipid	210*	1599	660
neutral lipid	62	493	189
phospholipid	138	1000	455
neutral glycolipid	1.7	28	2.7
ganglioside	7.5	78	13
molar ratio. total sphingoglycolipid to phospholipid	0.045	0.076	0.023

* Values expressed $\mu g/mg$ protein.

Abstracted from O. Renkonen, C. G. Gahmberg, K. Simons, L. Kääriäinen (1972).

5% of the total cell protein is in the plasma membrane, then one can calculate from this data that 50% of the cellular ganglioside of BHK21-C13 is surface located. Comparable data is as yet unavailable for 3T3 mouse cells although Yogeeswaran et al. (1972) have reported a 3–5 fold enrichment of the gangliosides in the surface membrane compared with whole cells. This was considered to be only a modest enrichment when compared to the marker enzymes whose specific activity increased 10–20 times.

A similar discrepancy between the enrichment of gangliosides in plasma membranes and the increase in specific activity of marker enzymes was reported in studies on rat liver and bovine mammary gland subcellular fractions (Keenan et al., 1972a, b). An 11.74 times enrichment in ganglioside in the plasma membrane was compared to a 48 fold increase in the specific activity of 5′ nucleotidase over the homogenate. In addition, purified rough endoplasmic reticulum (no data given) was also enriched in ganglioside, although mitochondria and nuclei showed the same levels as whole tissue. The authors calculated that about 25% of the total cellular ganglioside of rat liver is in the surface membrane.

In studies in our laboratory we have fractionated NIL2 cells labelled with $[1\text{-}^{14}C]$palmitate in an attempt to study subcellular distribution of the density dependent glycolipids (Table 2.10). Incorporation into plasma membrane glycolipids was consistently 7–8 times greater when compared to the cell homogenate on a protein basis. In samples of endoplasmic reticulum which contained very low levels of Na^+K^+ activated ATPase, there was still considerable labelling of glycolipid, although the specific activity was of the order of 3 times lower than in the plasma membrane fraction. Mitochondrial and nuclear fractions were not enriched in incorporation into glycolipid, and the label in nuclei could be considerably

Table 2.10

INCORPORATION OF (1-^{14}C) PALMITATE INTO THE SUBCELLULAR FRACTIONS OF DENSE NIL2 CELLS

Cells were labelled in log phase with (1-^{14}C) palmitate 0.25 μc/ml of growth medium and harvested by scraping 48 h later when confluent. Subcellular fractions were isolated and characterized from about 2.5 × 10^8 cells by the method of Graham (1972) involving nitrogen cavitation.

	Cell Homogenate	Nuclei	Mitochondrial	Plasma Membrane	Endoplasmic Reticulum
Na$^+$K$^+$ ATPase*	N.D.	N.D.	N.D.	19.0	< 1.0
NADH Diaphorase†	N.D.	N.D.	N.D.	< 10.0	105.0
Total incorporation per mg protein	0.71	0.46	0.56	3.25	1.1
Incorporation into Phospholipid/mg	0.56	0.33	0.45	2.53	0.86
Into Glycolipid/mg	0.05	0.03	0.05	0.39	0.12
Into neutral lipid/mg	0.10	0.09	0.05	0.33	0.13
% Distribution of label in Glycolipid					
CmonoH	8.3	9.4	5.9	9.7	5.3
CdiH	5.4	4.8	6.1	5.9	7.5
CtriH	19.0	19.8	17.9	21.0	21.5
CtetraH	34.4	32.8	31.0	30.6	35.2
CpentaH	16.3	18.5	17.9	18.3	19.8
GM$_3$	17.0	14.8	21.0	14.3	10.7

* μmoles ATP hydrolysed/h/mg protein.

† μmoles NADH oxidized/h/mg protein.

reduced by further purification. The density dependent glycolipids were not only found in the plasma membrane. The distribution of the label in the various glycolipids suggests that the subcellular fractions have a similar glycolipid pattern to whole cells (Klenk and Choppin, 1970, Yogeeswaran et al., 1972, Renkonen et al., 1972, in agreement with previous data on other cells. However, Weinstein et al. (1970) in a detailed analysis of the L-cell glycolipids found only two of the cellular glycolipids, hematoside and disialoganglioside, in the plasma membrane. Ceramide dihexoside and monosialoganglioside were suggested to be in intracellular structures. Glycolipids accounted for about 0.7% of total cell lipid and plasma membrane lipid indicating no enrichment over other lipid.

In a somewhat different approach, we added neuraminidase to monolayers of NIL2 cells prelabelled with [1-^{14}C]palmitate. A decrease in the ratio of label in GM$_3$/CdiH was observed which plateaued after 6–8 h (Fig. 2.4). Other neutral glycolipids were unaffected by the enzyme treatment as expected. At this point,

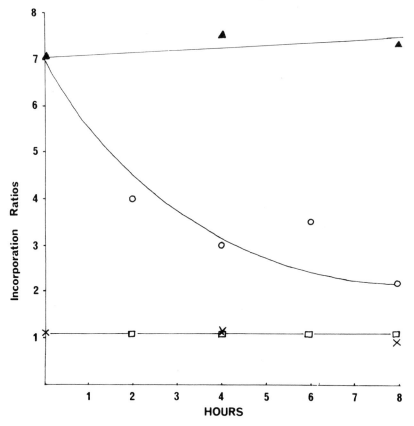

Fig. 2.4   The effect of neuraminidase on the glycolipids of NIL2 cells. NIL2 cells grown on 9 cm plastic petri dishes were labelled with 5 μc of $[1\text{-}^{14}C]$ palmitate for 24 h (cell no. 3×10^6/dish). Growth medium was removed and replaced with serum-free medium (6 ml) with or without 20 μg/ml neuraminidase (Cl. perfringens Type V1 1.3 units/mg. Sigma Chem. Co.). Throughout the incubation cells remained attached to the substratum; >90% of the cells excluded trypan blue after an 8 h incubation. Ratio of label in GM_3/CdiH, ▲–▲ control; ○–○ + enzyme; ratio of label in CtriH/CtetraH, ×–× control; □–□ + enzyme.

25% of the labelled cell hematoside had been converted to CdiH, suggesting that at least this amount was present in the surface membrane. Although 90% of the cells excluded trypan blue after the longest incubation period, the possibility that the enzyme slowly leaks into the cells cannot be discounted. Similar qualifications accompanied the finding that 50–60% of the lipid bound sialic acid of intact 3T3 cells was available to neuraminidase (Yogeeswaran *et al.*,1972). In contrast, Weinstein *et al.* (1970) reported that the two surface located gangliosides of L cells were not attacked by neuraminidase. A similar situation has been reported for the hematoside in cat erythrocytes (Wintzer and Uhlenbruck, 1967). In summary, the evidence supports the idea that a considerable part of the cellular glycolipids is in the surface membrane, although they may well be present in intra-cellular organelles at a lower concentration. The possibility remains that this is an artifact either due to cross contamination of subcellular fractions, or to redistribution between the plasma membrane and intracellular structures during homogenization (Wirtz and Zilversmidt, 1968). However, support for a genuine intracellular location comes from studies using fluorescent antibodies to glycolipids on intact tissues (Marcus and Janis, 1970).

The changes seen in cell glycolipid after transformation are thus potentially capable of influencing the properties of the cell surface. However, transformation of NIL2 cells for example, leads to a deletion of Ctri, tetra- and pentahexoside in endoplasmic reticulum and mitochondria as well as plasma membrane. That changes in cellular complex carbohydrates on transformation are not only located at the plasma membrane agrees with the data of Warren and colleagues (1972) concerning the distribution of a fucose containing glycoprotein characteristic of transformed cells.

REACTIVITY OF CELL BOUND GLYCOLIPID

There is good evidence that the absolute level of the glycolipid antigen in the cell membrane is not directly related to the reactivity of the component. Globoside for example is the main human erythrocyte glycolipid but anti-globoside serum does not react with the cell (Koscielak, Hakomori and Jeanloz, 1968). In contrast, the trace amounts of blood group A and B glycolipids are strongly antigenic and are highly reactive to their specific antisera. The globoside of the erythrocyte becomes highly reactive however if the cell is trypsinized. Foetal human erythrocytes are also highly reactive to antigloboside antiserum although the level in the cell is similar to that of the adult (Hakomori, 1971). The results suggest that during development, the reactive group of the glycolipid is becoming inaccessible or cryptic. In this context it was of great interest when Hakomori, Teather and Andrews (1968) found that viral transformation increased the reactivity of the hematoside in 3T3 and BHK21 cells to its antisera, although the chemical concentration decreased. Reactivity of normal cells was converted to that of the transformed by trypsinization. No data were presented on the reactivity of CdiH whose importance as an antigen in cancer tissue has already been demonstrated *in vivo*.

Similar experiments have been conducted with Forssmann antigen thought to be a glycolipid with N-acetylgalactosamine as the terminal determinant, although its structure is still a matter of controversy (Makita, Suzuki, Yosizawa, 1966; Siddiqui and Hakomori, 1971; Mallette and Rush, 1972). Burger (1971) reported that the low Forssmann activity of normal BHK cells in an haemolysis inhibition experiment could be activated to the high level found for the transformed cells, by trypsinization. The antigen in BHK was thought to contain lipid, which is not consistent with compositional data which shows no Forssmann type glycolipid in BHK (Hakomori, 1970b; D. Critchley, unpublished). Indeed, in experiments using NIL2 and polyoma transformed NIL2 cells and an antisera directed against purified Forssmann glycolipid, Hakomori and Kijimoto (1972) found the transformed cell barely reactive with the antisera (Fig. 2.5). Forssmann glycolipid synthesis is known to be activated at high cell density in NIL2 cells (Kijimoto and Hakomori, 1972),

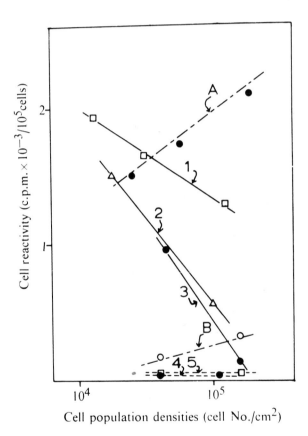

Cell population densities (cell No./cm^2)

Fig. 2.5   Reactivity and cell surface Forssmann antigen in normal and transformed NIL cells, and its dependence on cell density. Reactivity of intact cells was tested using anti-Forssman glycolipid rabbit antiserum labelled with ^{14}C-formaldehyde. Line A, reactivity of NIL 2K cells separated from cultures of different cell densities by 1% EDTA treatment. B, NIL2 E-Py similarly tested. Line 1, reactivity of NIL2 E cells tested in monolayer culture. NIL2 K cells, (lines 2 and 3) and NIL2 E-Py cells (lines 4, 5) were similarly tested.

Abstracted from S. Hakomori and S. Kijimoto (1972).

yet when tested in monolayer its reactivity decreased in dense culture. Confluent monolayers taken into suspension by EDTA were however more reactive than sparse cultures, again accentuating the idea that reactive groups can be masked, as has been shown many times in other systems. It would be interesting to know if the masking phenomenon is specific to glycolipids whose syntheses are density dependent. Reactivity of glycolipids has, as yet, not been studied in relation to the cell cycle.

RELATION OF GLYCOLIPID CHANGES TO TUMORIGENICITY OF THE CELL

The observation by Hakomori and Murakami (1968) that the glycolipids of a spontaneously tumorigenic BHK tended toward the pattern seen after transformation was not supported by the data of Brady and Mora (1970) in spontaneously tumorigenic Balb 3T3 and A/LN cells. We have asked similar questions about NIL2 cells the density-dependent glycolipids of which strikingly disappear on transformation (D. R. Critchley and I. Macpherson, 1973).

Tumours developed by injecting large numbers (10^7) of NIL2 cells into hamsters, were subsequently taken into culture and the glycolipid pattern analyzed. Of eleven tumour lines studied, ten had a significantly modified glycolipid pattern in that they no longer accumulated $[1-^{14}C]$palmitate into the Ctri and C tetrahexoside (Table 2.7). Incorporation into the ceramide pentahexoside remained, and was still density dependent. One of the tumour cell lines retained all three glycolipids in density dependent form. The pattern was only slightly modified on passage through an animal although viral transformation led to loss or marked reduction of the level of all three compounds.

Sakiyama and Robbins (1973) in a detailed study of the tumours induced by their NIL clones obtained essentially similar results. A NIL2 Cl cell line which contained density-dependent Ctri-, tetra-, and pentahexoside, produced tumours which on culturing contained only the pentahexoside. The component was still density dependent in two of three tumour cultures studied. A line NIL2 C2 in which pentahexoside was density dependent, produced tumours which in culture still contained the same component, with similar relation to culture density. The density dependence of GM_3 in NIL Cl was lost in the one tumour line studied. In addition, agglutinability of the clones by Concanavalin A or wheat germ agglutinin bore no clear correlation with tumorigenicity in contrast to the results of Inbar et al. (1972). It has also been reported that there was no correlation between the presence of normal gangliosides in a transformed 3T3 variant and agglutinability of the cell by Concanavalin A (Yogeeswaran et al., 1972).

DISCUSSION

The general concept that the malignant cell has a complement of glycolipids characterized by less complete carbohydrate chains, is supported by both *in vivo* and *in vitro* studies, although there are several exceptions. The change appears likely to affect the surface membrane, although similar modifications are probably reflected in intracellular glycolipids. The influence of glycolipids on cell surface properties has also been shown to be mediated by changes in their reactivity. The importance of both these mechanisms is apparent when considering the results on cytolipin H. That such a ubiquitous glycolipid should be an antigenic determinant in malignancy suggests that either it is present in much greater concentrations in tumours or it is "cryptic" in normal tissues, or a combination of both.

A similar theory for cell surface glycoproteins also exists, but mainly because of difficulties in characterizing these molecules, has yet to be adequately tested. However, the appearance of new or cryptic determinants in cancer cells may be related to such a phenomenon (Burger, 1971a, b; Black *et al.*, 1971). In this connection it was of great interest that the malignant cell could be "normalized" by coating with monovalent Concanavalin A, presumably through the blocking of reactive carbohydrate groups on the cell surface (Burger and Noonan, 1970). The density-dependent extension of carbohydrate chains of glycolipids may represent a similar phenomenon but in its physiological context. It may also be related to the proposal of Roseman and his colleagues (Roseman, 1970; Roth, McGuire and Roseman, 1971; Roth and White, 1972), that cellular recognition and adhesion are mediated by binding of surface glycosyl transferases on one cell to target molecules of substrate on the surface of contiguous cells. The CtriH formed in cultures of NIL2 and BHK21 cells in which there is appreciable cell contact may be synthesized by the process termed "transglycosylation". The accumulation of this glycolipid may thus be due to a sequence of reactions whose primary importance is cellular adhesion and recognition. It may have little direct function in growth regulation. The fact that transformed cells do not interact with each other in a normal way can be explained by the absence of the enzyme due to a specific block by the viral genome. An additional possibility is that the enzyme is inducible and the activity is low due to unavailability of substrate, perhaps due to the thicker surface coat characteristic of transformed cells (Martínez-Palomo, 1970). However, the fact that CtriH is found in the endoplasmic reticulum of cells suggests that this mode of synthesis is unlikely.

The synthesis of density dependent components, however, effected by cell contact, may lead to accumulation of, say, CtriH which in itself triggers at a critical level the transmission of a message to the nucleus leading to cessation of cell division. The lack of correlation between these glycolipids and saturation density or tumorigenicity may have several explanations. If such a chain of reactions is involved in regulation of cell division, a block at several points could result in the message going unnoticed. Thus the highly tumorigenic properties of NIL2/HSV may result from a block in synthesis of certain glycolipids, although the subsequent parts of the chain are in a functional state. The similar properties of tumour cell line 11 which has apparently normal glycolipids, Table 2.7, are explained by postulating that "mediators" between the cell surface and nucleus are defective. The failure of cyclic-AMP to restore normal glycolipids to transformed cells (Sakiyama *et al.*, 1972; Yogeeswaran *et al.*, 1972) which have recovered some properties of growth regulation, is perhaps because cyclic AMP is one such mediator, and thus by-passes the need for a glycolipid trigger.

The lack of absolute correlation between tumorigenicity and glycolipids may on the other hand, reflect the presence of much stronger determinants of cell surface properties in other molecules. The fact that density dependent glycolipids have only been found in a few cell lines makes it desirable to look for a similar glycosyl extension response on glycoproteins.

The concentration in this discussion on the glycosyl extension response does not invalidate the more general concept that the glycolipids characteristic of normal cells, density dependent or not, are in some way involved in cellular recognition and growth regulation, perhaps involving the binding of target molecules to receptor proteins. It may be the reduction in concentration of these molecules at the transformed cell surface which is an important factor in the behaviour of these cells. Proof of such a key role in control of cellular events necessitates a more direct approach, perhaps employing glycosidases or glycolipid antibodies to modify the normal cell surface.

REFERENCES

Adams, E. P. and Gray, G. M. (1967). *Nature, 216*, 277.

Black, P. H., Collins, J. J. and Culp, L. A. (1971). In Oncology, 1970 Proc. Tenth Int. Cancer Cong., Vol. 1, R. L. Clark, R. W. Cumley, J. E. McCoy and M. M. Copeland, eds., Year Book Medical Publishers Inc., Chicago, Ill., p. 210.

Brady, R. O. (1970). *Chem. Phys. Lipids, 5*, 261.

Brady, R. O., Borek, C. and Bradley, R. M. (1969). *J. Biol. Chem., 244*, 6552.

Brady, R. O. and Mora, P. T. (1970). *Biochem. Biophys. Acta.*, *218*, 308.

Burger, M. M. (1971a). *Nature New Biology*, *231*, 125.

Burger, M. M. (1971b). In L. A. Manson, *Biomembranes*, Vol. 2, Plenum Press, New York, p. 247.

Burger, M. M. and Noonan, K. D. (1970). *Nature*, *228*, 512.

Cheema, P., Yogeeswaran, G., Morris, H. P. and Murray, R. K. (1970). *Febs. Letters*, *11*, 181.

Coles, L., Hay, J. B. and Gray, G. M. (1970). *J. Lipid Res.*, *11*, 158.

Critchley, D. R. and Macpherson, I. (1973). *Biochim. Biophys. Acta. 296*, 145.

Cumar, F. A., Brady, R. O., Kolodny, E. H., McFarland, V. W. and Mora, P. T. *Proc. Nat. Acad. Sci.*, U.S.A., *67*, 757.

Dawson, G., Matalon, R. and Dorfman, A. (1972). *J. Biol. Chem.*, *247*, 5944.

Den, H., Shultz, A. M., Basu, M. and Roseman, S. (1971). *J. Biol. Chem.*, *246*, 2721.

Dijong, I., Mora, P. T. and Brady, R. O. (1971). *Biochem.*, *10*, 4039.

Dod, B. J. and Gray, G. M. (1968a). *Biochem. J.*, *110*, p. 50p.

Dod, B. J. and Gray, G. M. (1968b). *Biochim. Biophys. Acta.*, *150*, 397.

Eckhart, W., Dulbecco, R. and Burger, M. M. (1971). *Proc. Nat. Acad. Sci.*, U.S.A., *68*, 283.

Fishman, P. H., McFarland, V. W., Mora, P. T. and Brady, R. O. (1972). *Biochem. Biophys. Res. Comm.*, *48*, 48.

Graham, J. M., *Biochem. J.* (1973).

Gray, G. M. (1971). *Biochim. Biophys. Acta, 239*, 494.

Hakomori, S. (1970a). *Chem. Phys. Lipids*, *5*, 96.

Hakomori, S. (1970b). *Proc. Nat. Acad. Sci.*, U.S.A., *67*, 1741.

Hakomori, S. (1971). In D. F. H. Wallach and H. Fischer, Dynamic Structure of Cell Membranes, Springer Verlag, Berlin, p. 65.

Hakomori, S. and Kijimoto, S. (1972). *Nature New Biology*, *239*, 87.

Hakomori, S., Kijimoto, S. and Siddiqui, B. (1971a). *Fed. Proc.*, *30*, Abstract 1043.

Hakomori, S., Koscielak, J., Bloch, K. J. and Jeanloz, R. W. (1967). *J. Immunol.*, *98*, 31.

Hakomori, S. and Murakami, W. T. (1968). *Proc. Nat. Acad. Sci.*, U.S.A., *59*, 254.

Hakomori, S., Saito, T. and Vogt, P. K. (1971b). *Virology*, *44*, 609.

Hakomori, S., Teather, C. and Andrews, H. (1968). *Biochim. Biophys. Res. Comm.*, *33*, 563.

Howard, B. V. and Kritchevsky, D. (1969). *Biochim. Biophys. Acta, 187*, 293.

Inbar, M., Ben-Bassat, H. and Sachs, L. (1972). *Nature New Biology*, *236*, 3.

Kawanami, J. and Tsuji, T. (1968). *Jap. J. Exp. Med.*, *38*, 11.

Keenan, T. W., Huang, C. M. and Morré, D. J. (1972). *Biochem. Biophys. Res. Commun.*, *47*, 1277.

Keenan, T. W., Morré, D. J. and Huang, C. M. (1972). *Febs. Letters*, *24*, 204.

Kijimoto, S. and Hakomori, S. (1971). *Biochem. Biophys. Res. Comm.*, *44*, 557.

Kijomoto, S. and Hakamori, S. (1972). *Febs. Letters*, *25*, 38.

Klenk, H. D. and Choppin, P. W. (1970). *Proc. Nat. Acad. Sci.*, U.S.A., *66*, 57.

Koscielak, J., Hakomori, S. and Jeanloz, R. W. (1968). *Immunochemistry*, *5*, 441.

Kraemer, P. M. (1971). In L. A. Manson, *Biomembranes*, *1*, Plenum Press, New York, p. 67.

Losick, R. M. and Robbins, P. W. (1969). *Sci. Am.*, *221*, 121.

Makita, A., Suzuki, C. and Yosizawa, Z. (1966). *J. Biochem.*, *60*, 502.

Mallette, M. F. and Rush, R. L. (1972). *Immunochemistry*, *9*, 809.

Marcus, D. M. and Cass, L. E. (1969). *Science*, *164*, 553.

Marcus, D. M. and Janis, R. (1970). *J. Immunol.*, *104*, 1530.

Martensson, E. (1969). In R. T. Holman, Progress in the Chemistry of Fats and Other Lipids, Vol. 10, Pergamon Press, New York, p. 367.

Martin, G. S. (1970). *Nature*, *227*, 1021.

Martin, G. S., Venuta, S., Weber, M. and Rubin, H. (1971). *Proc. Nat. Acad. Sci.*, *68*, 2739.

Martinez-Palomo, A. (1970). *Int. Rev. Cytol.*, *29*, 29.

Mora, P. T., Brady, R. O., Bradley, R. M. and McFarland, V. W. (1969). *Proc. Nat. Acad. Sci.*, U.S.A., *63*, 1290.

Mora, P. T., Cumar, F. A. and Brady, R. O. (1971). *Virology*, *46*, 60.

Morell, P. and Braun, P. (1972). *J. Lipid Res.*, *13*, 293.

Pollack, R. E., Green, H. and Todaro, G. J. (1968). *Proc. Nat. Acad. Sci.*, U.S.A., *60*, 126.

Rapport, M. M. (1969). *Ann. N.Y. Acad. Sci.*, *159*, 446

Rapport, M. M., Graf, L., Skipski, V. P. and Alonzo, N. F. (1958). *Nature*, *181*, 1803.

Rapport, M. M. and Graf, L. (1969). In P. Kallos and B. H. Waksman, Progress in Allergy, Vol. 13, S. Karger, Basel, p. 273.

Rapport, M. M., Schneider, H. and Graf, L. (1967). *Biochim. Biophys. Acta*, *137*, 409.

Renkonen, O., Gahmberg, C. G., Simons, K. and Kääriäinen, L. (1970). *Acta. Chem. Scand.*, *24*, 733.

Renkonen, O., Gahmberg, C. G., Simons, K. and Kääriäinen, L. (1972). *Biochim. Biophys. Acta*, *255*, 66.

Robbins, P. W. and Macpherson, I. A. (1971a). *Proc. Roy. Soc.*, London (Biol.), *177*, 49.

Robbins, P. W. and Macpherson, I. (1971b). *Nature*, *229*, 569.

Roseman, S. (1970). *Chem. Phys. Lipids*, *5*, 270.

Roth, S., McGuire, E. J. and Roseman, S. (1971). *J. Cell Biol.*, *51*, 536.

Roth, S. and White, D. (1972). *Proc. Nat. Acad. Sci.*, U.S.A., *69*, 485.

G. H. Rothblat and D. Kritchevsky (Eds). Lipid Metabolism in Tissue Culture Cells (1966). Wistar Institute Symposium Monograph No. 6.

Sakiyama, H., Gross, S. R. and Robbins, P. W. (1972). *Proc. Nat. Acad. Sci.*, U.S.A., *69*, 872.

Sakiyama, H. and Robbins, P. W. (1973). *Fed. Proc.*, *32*, 86.

Siddiqui, B. and Hakomori, S. (1970). *Cancer Res.*, *30*, 2930.

Siddiqui, B. and Hakomori, S. (1971). *J. Biol. Chem.*, *246*, 5766.

Siefert, H. and Uhlenbruk, G. (1965). *Naturwissenshaften*, *32*, 190.

Svennerholm, L. (1964). *J. Neurochem.*, *11*, 839.

Svennerholm, L. (1970). In M. Florkin and E. H. Stotz, Comprehensive Biochemistry, Vol. 18, Elsevier Publishing Co., Amsterdam, p. 201.

Sweeley, C. C. and Dawson, G. (1969). In G. A. Jamieson and T. J. Greenwalt, *Red Cell Membrane*, Lippincott Co., Philadelphia, p. 172.

Tal, C. (1965). *Proc. Nat. Acad. Sci.*, U.S.A., *54*, 1318.

Tal, C., Dishon, T. and Gross, J. (1964). *Brit. J. Cancer*, *18*, 111.

Tal, C. and Halperin, M. (1970). *Israel J. Med. Sci.*, *6*, 708.

Warren, L., Critchley, D. and Macpherson, I. (1972). *Nature*, *235*, 275.

Warren, L., Fuhrer, J. P. and Buck, C. A. (1972). *Proc. Nat. Acad. Sci.*, U.S.A., *69*, 1838.

Weigandt, H. (1967). *J..Neurochem.*, *14*, 671.

Weinstein, D. B., Marsh, J. B., Glick, M. C. and Warren, L. (1970). *J. Biol. Chem.*, *245*, 3928.

Wiblin, C. N. and Macpherson, I. A. (1972). *Int. J. Cancer*, *10*, 296.

Wintzer, G. and Uhlenbruck, G. (1967). *Z. Immunitatsforsch. Allerg.*, *133*, 60.

Wirtz, K. W. A. and Zilversmit, D. B. (1968). *J. Biol. Chem.*, *243*, 3596.

Yang, H. and Hakomori, S. (1971). *J. Biol. Chem.*, *246*, 1192.

Yogeeswaran, G., Sheinin, R., Wherrett, J. R. and Murray, R. K. (1972). *J. Biol. Chem.*, *247*, 5146.

Yogeeswaran, G., Wherrett, J. R., Chatterjee, S. and Murray, R. K. (1970). *J. Biol. Chem.*, *245*, 6718.

3 Glycolipids in Malignancy. Alterations in Membrane Glycolipids in Tumorigenic Virus-Transformed Cell Lines

Roscoe O. Brady and Peter H. Fishman

Developmental and Metabolic Neurology Branch, National Institute of Neurological Diseases and Stroke, National Institutes of Health, Bethesda, Maryland

The loss of contact inhibition of growth exhibited by transformed cells in culture is widely used at the present time as a model for investigating carcinogenesis. In the past, the phenomenon of contact inhibition had not been associated with definitive chemical properties of the cell surface. However, within the past few years, some indications of the components involved have appeared. For example, neoplastic cells exhibit tumor specific surface antigens (Tevethia *et al.*, 1965; Habel, 1967; Smith *et al.*, 1970), changes in agglutinability by certain plant lectins (Burger and Goldberg, 1967; Inbar *et al.*, 1969), alteration in glycolipids (Hakomori and Murakami, 1968; Mora *et al.*, 1969; Robbins and MacPherson, 1971), and glycoproteins (Wu *et al.*, 1969; Buck *et al.*, 1970; Sakiyama and Burge, 1972).

Because we had observed changes in the activity of certain sphingolipid hydrolases in leukemic leukocytes (Kampine *et al.*, 1967) and acidic glycolipids known as gangliosides are highly concentrated in membranous elements of cells, we undertook an examination of the composition and metabolism of these complex lipids in virus-transformed cells. Our work has largely dealt with studies of gangliosides in established mouse cell lines for the following reasons: (1) Cultured mouse cells contain the full complement of the types and distribution of gangliosides usually found in most mammalian tissues (Fig. 3.1). (2) Many different mouse cell lines with spontaneous and virally transformed clonal derivatives are

Ceramide = N-fatty acyl sphingosine

$$CH_3-(CH_2)_{12}-CH=CH-CH(CH)-CH-C*H_2OH$$

$$\begin{array}{c} | \\ N-H \\ | \\ C=O \\ | \\ (CH_2)_{16} \\ | \\ CH_3 \end{array}$$

Hematoside (G_{M3}): Ceramide-glucose-galactose-N-acetylneuraminic acid

(G_{M2}): Ceramide-glucose-galactose-N-acetylgalactosamine
$$\begin{array}{c} | \\ \text{N-Acetylneuraminic acid} \end{array}$$

(G_{M1}): Ceramide-glucose-galactose-N-acetylgalactosamine-galactose
$$\begin{array}{c} | \\ \text{N-Acetylneuraminic acid} \end{array}$$

(G_{D1a}): Ceramide-glucose-galactose-N-acetylgalactosamine-galactose
$$\begin{array}{cc} | & | \\ \text{N-Acetylneuraminic acid} & \text{N-Acetylneuraminic acid} \end{array}$$

* Indicates point of attachments of various substituents.

Fig. 3.1 Structures of the principal gangliosides of contact inhibited mouse cell lines.

available for examination. (3) The use of cells well established in culture diminishes the likelihood of non-specific changes which might occur as primary explants and secondary cultures undergo selection and population drift. (4) The genotypic and phenotypic properties of normal and transformed cell lines are well defined. (5) Cell lines derived from highly inbred mice are available which are eminently suitable for determining the tumorigenicity of such cells.

There is a striking difference in the ganglioside pattern of Simian virus 40 transformed cells (SVS AL/N) compared with that of normal (N AL/N) or spontaneously transformed cells (TAL/N) (Fig. 3.2). Gangliosides in these cells appear as doublets on thin-layer chromatograms. The physical chemical reason for this behavior is due to the presence of the different analogs of sialic acid, N-acetylneuraminic and N-glycolylneuraminic acid; gangliosides with the latter derivative migrate slightly slower. Additionally, the presence of different fatty acids in the ceramide moiety of the ganglioside can cause differences in migration rate. A similar alteration in ganglioside composition was observed in a polyoma virus-transformed AL/N cell line (PY AL/N) and in DNA virus-transformed Swiss mouse 3T3 cell lines (Brady and Mora, 1970) Table 3.1. This difference in ganglioside pattern was also seen in an SV40

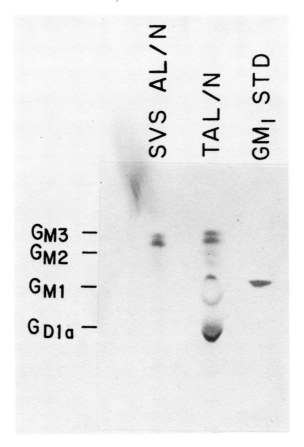

Fig. 3.2 Thin-layer chromatogram of gangliosides of mouse AL/N cell lines. T AL/N is a spontaneously transformed line; SVS AL/N is a SV40 transformant. After the gangliosides were isolated from the respective types of cells, they were separated by thin-layer chromatography and visualized with resorcinol spray agent. The ganglioside pattern of T AL/N cells is essentially like that of N AL/N (Brady and Mora, 1970). Reproduced with permission from Mora *et al.*, 1969.

transformed clonal Balb/c cell line (Brady and Mora, 1970). Such changes were not observed in spontaneously transformed TAL/N or Balb/c 3T12 cells nor tumor derived (T AL/NT) cells when recultured *in vitro*. The difference in ganglioside distribution was further substantiated by gas-liquid chromatographic analysis of the isolated gangliosides (Dijong *et al.*, 1971) and the findings are summarized in schematic form in Fig. 3.3.

Table 3.1

DISTRIBUTION OF GANGLIOSIDES IN AL/N AND SWISS MOUSE
CELL LINES AND IN SV40 AND POLYOMA VIRUS-TRANSFORMED
DERIVATIVE LINES

| Cell Type | Gangliosides | | | | |
	G_{D1a}	G_{M1}	G_{M2}	G_{M3}	Total
	nmoles/mg of protein				
N AL/N	1.8	1.5	0.8	0.6	4.7
SVS AL/N	0.16	0.22	0.1	1.9	2.4
PY AL/N	0.1	0.15	0.2	1.8	2.3
Swiss 3T3	2.4	2.6	1.8	4.0	10.8
PY 11*	0.1	0.2	0.05	3.2	3.5
SV 101†	0.05	0.1	0.05	3.5	3.7

* Polyoma virus transformed Swiss mouse cell line.

† Simian virus 40 transformed Swiss mouse cell line. From Mora *et al.* (1969)
and Brady and Mora (1970).

GANGLIOSIDES OF NORMAL OR TUMORIGENIC
MOUSE CELL LINES

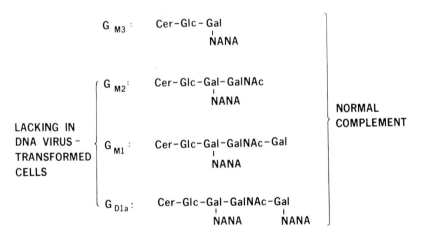

Fig. 3.3 Gangliosides of normal or tumorigenic mouse cell lines.

METABOLIC STUDIES

The absence of gangliosides larger than G_{M3} in the virally
transformed cells is also reflected in biosynthetic studies. If N-acetyl-
[^{3}H]-mannosamine, a precursor of sialic acid, is added to the culture
media, radioisotope is primarily incorporated into G_{D1a} in the

normal contact-inhibited cells and into G_{M3} in the SV40-trans-
formed (Fig. 3.4). The alteration in ganglioside distribution could
conceivably be due to either increased catabolism or decreased

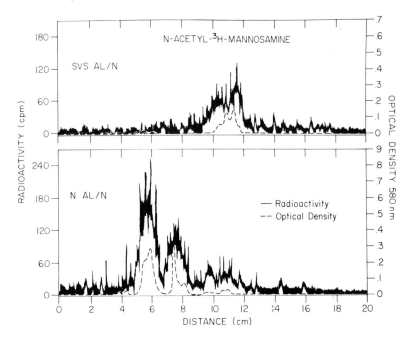

Fig. 3.4 N AL/N and SVS AL/N cells were grown in the presence of N-acetyl-
[³H] mannosamine. The gangliosides were isolated and separated by thin-layer
chromatography. After the distribution of labelled gangliosides was determined
by scanning the plates for radioactivity, the gangliosides detected by spraying
resorcinol. Reproduced with permission from Brady and Mora, 1970.

synthesis of the higher ganglioside homologs. The major ganglioside
in normal mouse cells is G_{D1a} and in virally transformed cells, G_{M3}.
Both of these glycolipids have a terminal molecule of sialic acid and
their catabolism is initiated through the activity of a neuraminidase
on the respective compounds. We investigated these initial hydroly-
tic reactions in contact-inhibited and virus-transformed cells and
found that the activity of these catabolic enzymes was essentially the
same in these preparations (Cumar *et al.*, 1970).

We then undertook an investigation of specific reactions
involved in the biosynthesis of gangliosides in the different cell lines.
The enzymatic synthesis of gangliosides proceeds by the sequential
addition of sugar moieties from sugar nucleotide donors to the
growing glycolipid acceptor. As G_{M3} is present in the virally transformed cell lines but more complex homologs are virtually absent,

the activity of the N-acetylgalactosaminyltransferase involved in the synthesis of G_{M2} from G_{M3} was investigated (Reaction 1):

1. Cer-Glc-Gal-NANA (G_{M3}) + UDP-GalNAc $\xrightarrow[\text{galactosaminyltransferase}]{\text{hematoside : N-acetyl-}}$

$$\text{Cer-GlC-Gal(NANA)-GalNAc } (G_{M2}) + \text{UDP}$$

This enzyme is readily detected in homogenates of normal and spontaneously transformed mouse cell lines and it appears to be localized in membrane-rich subcellular particles. The activity of this N-acetylgalactosaminyltransferase is drastically reduced in SV40 and polyoma virus-transformed cells (Cumar *et al.*, 1970) (Table 3.2).

Table 3.2
URIDINE DIPHOSPHATE N-ACETYLGALACTOSAMINE:
HEMATOSIDE N-ACETYLGALACTOSAMINYLTRANSFERASE
ACTIVITY IN MOUSE CELL LINES

Cell Line	Glycolipid Acceptor	
	$G_{M3}NAc*$	$G_{M3}NGlyc\dagger$
	(Activity as percent of control cells)	
N AL/N	100	100
T AL/N	189	100
SVS AL/N	0	10
PY AL/N	0	0
Swiss 3T3	100	100
SV 101	33	30
PY 11	26	2

* N-acetylneuraminylgalactosylglucosylceramide (hematoside).
† N-glycolylneuraminylgalactosylglucosylceramide (hematoside).
Data calculated from Cumar *et al.* (1970).

Effect of Growth and Culture Conditions on Ganglioside Metabolism

Normal cells in culture will cease dividing when they come into contact with one another. Transformed cells, as opposed to contact-inhibited cells, continue to grow and divide after making contact and will reach a high saturation density in culture. Usually the ganglioside composition and glycosyltransferase activities were determined when both types of cells were subconfluent and still growing exponentially. The possible influence of cell density on the ganglioside content and ganglioside biosynthetic enzymes was carefully explored with Swiss 3T3 cells. There was no apparent effect over a ten-fold range of cell density on the ganglioside pattern nor a significant change in the activities of glycosyltransferases (Fishman *et al.*, 1972b). Similar results were obtained with a contact-inhibited Balb 3T3 cell line, two SV40-transformed Balb 3T3 cell

lines, and a flat revertant SV40 transformed Balb 3T3 cell line.
The amino sugar transferase activity was unaffected by sparse or
confluent culture conditions.

Since the spontaneously transformed cell lines have G_{M3}:N-
acetylgalactosaminyltransferase activity and grow as fast or faster
than the virally transformed cells and to high saturation density,
these growth properties cannot be a factor in the reduced aminosugar transferase activity found in the virally transformed lines.
Also the number of generations in culture appears not to be a
factor as the spontaneously transformed lines have been in culture
over 200 generations and still have high N-acetylgalactosaminyltransferase activity. However, extended cultivation of spontaneously
transformed lines can result in other changes in ganglioside metabolism. Both Balb 3T12 and T AL/N cells after 200 passages in
culture exhibited an altered ganglioside distribution characterized
by an absence of G_{D1a} and G_{M1} and increased G_{M2} and G_{M3} while
T AL/N by passage 150 had increased levels of G_{M2}. Ganglioside
synthesizing enzymes were assayed on early and late passage
T AL/N cells. Galactosyltransferase activity (Reaction 2) was very
low in the high passage cells; other enzymes involved in ganglioside
synthesis were within normal range (Fishman, Brady and Mora,
1972a).

2. Cer-Glc-Gal-(NANA)-GalNAc (G_{M2}) + UDP-Gal $\xrightarrow{\text{$G_{M2}$:galactosyl-}\atop\text{transferase}}$

$\qquad$ Cer-Glc-Gal-(NANA)-GalNAc-Gal(G_{M1}) + UDP

RELATIONSHIP BETWEEN VIRAL TRANSFORMATION
AND ALTERED GANGLIOSIDE METABOLISM

These metabolic studies suggest an alteration in ganglioside
metabolism as an expression of a virus-regulated or mediated
biochemical function common to both SV40 and polyoma induced
transformation of the mouse cell lines. One could at this time pose
several questions.

A. *Lytic Infection of Mouse Cells*

If the changes observed are related to some viral function,
is it a function involved in transformation and/or productive
infection? Mouse cells are permissive for polyoma virus which will
lytically infect these cells and only rarely transform them (cf.
Eckhart, 1969). Both T AL/N and Swiss 3T3 cells were treated with
a high multiplicity of polyoma virus to ensure complete infection of
all of the cells. There is an induction of cellular DNA synthesis and
of enzyme activities involved in DNA synthesis upon productive
infection with oncogenic DNA viruses, which peaks at around

28 h (Dulbecco *et al.*, 1965). At 50 h, virion protein and complete but unreleased virions are present in the cells. N-acetylgalactosaminyltransferase activity was assayed at these two times and the levels were similar to that in the uninfected controls (Mora, Cumar and Brady, 1971). The cells underwent lysis at 4–7 days, a sign of successful lytic infection. The results indicate that the reduced transferase activity in virally transformed cells is specifically associated with some transforming function of these DNA viruses.

B. *Phenotypic Reversion of Growth Properties and Altered Ganglioside Metabolism*

Since the change in ganglioside biosynthesis appears to be related to transformation, is the presence of an integrated viral genome sufficient to produce this change? Phenotypic "flat revertant" cloned lines derived from SV40 and polyoma transformed Swiss 3T3 cells (Pollack *et al.*, 1968) were examined for ganglioside composition. In the "flat revertant" SV40 line (F1^2 SV101) which is now contact inhibited, the ganglioside pattern was similar to the normal contact inhibited Swiss 3T3 cell line as opposed to the non-contact inhbited SV40 transformed cells (Table 3.3). In the "flat

Table 3.3
DISTRIBUTION OF GANGLIOSIDES IN SWISS 3T3 CELL LINE, IN SV40 AND POLYOMA VIRUS-TRANSFORMED DERIVATIVE LINES, AND IN FLAT SUBLINES DERIVED FROM THE VIRUS-TRANSFORMED LINES

| Cell Lines | Gangliosides (nmoles/mg protein) | | | | |
	G_{D1a}	G_{M1}	G_{M2}	G_{M3}	Total
Swiss 3T3	1.8	1.4	1.3	3.3	7.8
SV101	0.05	0.1	0.05	3.5	3.7
PY11	0.1	0.2	0.05	3.2	3.5
F1^2 SV101	2.6	0.8	1.4	2.9	7.7
F1 PY11	0.93	0.2	0.4	3.9	5.4

From Mora, Cumar and Brady (1971).

revertant" cell line derived from polyoma virus-transformed cells (F1 PY11) the cells showed phenotypic and genotypic heterogeneity (Pollack *et al.*, 1970) and the ganglioside pattern was only partially restored to normal. The level of N-acetylgalactosaminyltransferase activity is fully restored in F1^2 SV101 cells and partially restored in the F1 PY11 cell line (Table 3.4). Similar results were obtained with a "flat" clonal cell line selected out from SV40 transformed Balb 3T3 mouse cells by a serum factor growth dependence (cf. Fishman, Brady and Mora, 1972a).

Table 3.4
URIDINE DIPHOSPHATE N-ACETYLGALACTOSAMINE:
HEMATOSIDE N-ACETYLGALACTOSAMINYL-TRANSFERASE
ACTIVITY IN SWISS MOUSE CELL LINES

| Cells | Glycolipid Acceptor | |
	$G_{M3}NAc$	$G_{M3}NGlyc$
	Transferase Activity (Percent of Controls)	
3T3	100	100
SV 101	33	30
$F1^2SV$ 101	202	128
PY11	26	2
F1 PY11	46	22

From Mora, Cumar and Brady (1971).

The results indicate that the altered ganglioside metabolism as manifested by the activity of the N-acetylgalactosaminyltransferase is coordinately linked to the phenotypic growth properties of these virally transformed mouse cells. Thus, a biochemical expression of transformation appears to be coupled to the morphological consequence of transformation. Since in these "flat revertant" cell lines the tumor virus is integrated into the host cell DNA, T-antigen is produced which indicates expression of some viral properties and, in the case of SV40, virus can be rescued from these cells, it appears that the expression of certain viral functions can be modulated by the host cell. Thus, integration of the viral genome into host cell DNA in itself is not sufficient to cause altered ganglioside metabolism. The point is further illustrated by studies with an "abortive" SV40 transformed Balb 3T3 clonal line. This cell line does not express any viral functions but contains 3 to 6 viral genome equivalents as detected by DNA-DNA hybridization experiments (Smith et al., 1972). There was no decrease of N-acetylgalactosaminyltransferase in these cells when compared to the parental line (Fishman, Brady and Mora, 1972a).

C. Stability of the Altered Ganglioside Metabolism in DNA Virus-Transformed Mouse Cells

The reduction in the content of higher gangliosides and in the activity of the aminosugar transf   observed after oncogenic DNA virus-transformation appears to be a stable inherited characteristic of these transformed mouse cell lines. Several of the lines have been in continuous culture for over two years (over two hundred passages) and still exhibit an abnormal ganglioside pattern and low enzyme activity. In addition, SV40 transformed cells have been injected into syngenetic immunosuppressed mice and the induced tumors passed back into tissue culture. Even after three repetitions of this procedure, the tumor derived cell lines in culture still maintained low levels of N-acetylgalactosaminyltransferase (unpublished experiments of F. A. Cumar, R. W. Smith and P. T. Mora). In contrast T AL/N cells carried through the same procedure retained their high enzyme activity (unpublished results of P. H. Fishman and R. W. Smith). These experiments clearly demonstrate that this property of tumor virus transformed cells is very stable and can withstand very strong selective pressures.

D. *Specificity of the Change in Ganglioside Metabolism*

Another important question is how specific is the effect? Since the ganglioside biosynthetic enzymes apparently exist as a multienzyme complex, are other ganglioside glycosyltransferases altered by viral transformation? The levels of four of these transferases (Fig. 3.5) were determined in normal and virally transformed Swiss 3T3 cells. The SV40 transformant had decreased levels of sialyltransferase II (up to three-fold) and N-acetylgalactosaminyltransferase (up to eight fold) but the same or elevated levels of sialyltransferase I and galactosyltransferase (Table 3.5). The levels in the polyoma transformant were similar to the normal Swiss 3T3 cell line except for the aminosugar transferase (reduced seven-fold). Similar experiments were undertaken with AL/N cells. Sialyltransferase I is lower in SV40 and higher in polyoma transformed cells while sialyltransferase II is elevated in both transformed lines (Table 3.6). Again the N-acetylgalactosaminyltransferase levels are dramatically down in both transformants.

Sialyltransferase I was reported to be lower in a polyoma transformed baby hamster kidney (BHK) cell line (H. Den *et al.*, 1971). Reduced sialyltransferase II was also observed in SV40 transformed Swiss 3T3 cells (Grimes *et al.*, 1971), but not in polyoma transformed Swiss 3T3 (Grimes, personal communication). The low galactosyltransferase in the virally-transformed AL/N cell lines remains unexplained. However, these cell lines have undergone extensive cultivation and as mentioned earlier, the spontaneously transformed T AL/N cell line after prolonged passage in culture have very low levels of galactosyltransferase activity compared to early passage T AL/N or NAL/N cells. Otherwise, except for hematoside: N-acetylgalactosaminyl transferase, other transferase activities were the same as in the controls or even elevated.

The alterations in ganglioside biosynthesis after transformation of established mouse cells by these two oncogenic DNA viruses are complex. However, the one consistent change in both cell types

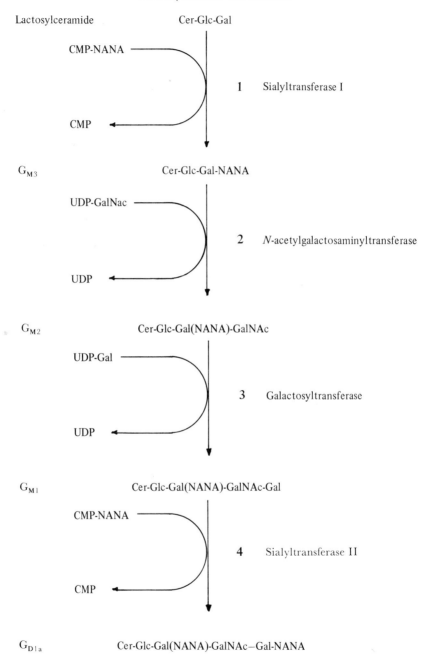

Fig. 3.5   Schematic representation of the steps involved in the enzymatic synthesis of gangliosides.

Table 3.5

GLYCOSYLTRANSFERASE ACTIVITIES IN NORMAL AND
VIRALLY-TRANSFORMED SWISS 3T3 CELLS

	Cell Line	
Glycosyltransferase	SV 101	PY 11
	(Percent of that in contact inhibited 3T3 cells).	
Sialyltransferase I	105	110
N-acetylgalactosaminyltransferase	13	17
Galactosyltransferase	655	168
Sialyltransferase II	36	82

Data from Fishman *et al.* (1972).

Table 3.6

GLYCOSYLTRANSFERASE ACTIVITIES IN NORMAL AND
VIRALLY-TRANSFORMED AL/N MOUSE CELLS

	Cell Line	
Glycosyltransferase	SVS-AL/N	PY-AL/N
	Percent of that in contact inhibited N AL/N cells	
Sialyltransferase I	56	143
N-acetylgalactosaminyltransferase	16	14
Galactosyltransferase	6	11
Sialyltransferase II	184	192

Data from Fishman *et al.* (1972).

after transformation by either SV40 or polyoma is the marked
reduction in N-acetylgalactosaminyltransferase activity. This observation supports the concept of a common biochemical change in
these virally transformed cells (Mora, Cumar and Brady, 1971).

E. *Generality of the Phenomenon*

A final and most important question is whether the observed
change in ganglioside metabolism is a general phenomenon of viral
transformation. One of the great difficulties in investigating changes
associated with cell transformation by oncogenic DNA viruses in the
low efficiency of transformation. Only a small percentage of the cell
population is stably transformed by the virus and the transformants
must be selected out and grown for several generations before sufficient quantities of cells are available for biochemical analyses.

It is therefore, crucial to determine whether the observed
changes are related to viral transformation or due to a coincidental
selection of those few mouse cells which may have a low level of

aminosugar transferase. There is a wide variation in the activity of this enzyme not only between different mouse cell strains but between different clones of the same strain. However, this enzyme activity has consistently been found to be lower in SV40 or polyoma transformed clonal derivatives when compared to the parental cell line. To date our group has examined over twenty virally transformed mouse cell lines and found this difference in all but one SV40 transformed mouse cell line. This line had been selected for 5-bromo-2-deoxyuridine resistance (Dubbs et al., 1967).

Similar changes in ganglioside distribution in SV40 and polyoma virus-transformed Swiss 3T3 cells have been observed independently (Sheinin et al., 1971). However, this same laboratory has recently reported that two SV40 transformed clones of mouse 3T3 cells have increased G_{D1a} (Yogeeswaran et al., 1972). In order to answer this important question, we are in the process of investigating ganglioside metabolism in a number of clonal derivatives of recently transformed SV40 mouse cell lines.

Currently, we are examining the effects on ganglioside metabolism when Swiss 3T3 cells are transformed by the RNA tumor virus, murine sarcoma virus. Since mass transformation of cell population occurs with this virus, the problems of selection are avoided. In addition, the kinetics of transformation and altered ganglioside metabolism can be followed. The results indicate an altered ganglioside metabolism similar to that observed with DNA virus transformation (Fishman et al., 1972c).

MOLECULAR BASIS OF ALTERED GANGLIOSIDE METABOLISM

By what mechanism can the insertion of a tumor virus genome into host cell DNA affect the activity of a host cell enzyme? Some preliminary experiments were undertaken to explore this area. When homogenates or subcellular fractions of virally transformed cells and non-virally transformed cells were mixed together and assayed for aminosugar transferase activity there was no evidence of an enzyme activator in the non-virus-transformed line (Mora, Cumar and Brady, 1971). Furthermore, when the above cell types were grown under common media, there was no evidence of a diffusible inhibitory factor. However, when the two cell types were grown in a mixed culture, there was a depression of aminosugar transferase activity. The possibility of some regulatory substance being transferred from virus-transformed cells to control cells upon intimate contact will have to be further explored.

One possible mechanism would be for the two viruses SV40 and polyoma to code for a common repressor. Both viruses code for 5–10 polypeptides (Eckhart, 1969), one of which might be a repressor of the aminosugar transferase. Recent studies with temperature-sensitive mutants of polyoma virus indicate that at least one viral gene product is necessary for transformation. Experiments with temperature-sensitive virus-transformed mouse cells may provide an answer to this question.

Alternatively, integration of the viral genome into host cell DNA may occur at the site of the gene coding for aminosugar transferase. This could result in a blocking of transcription or the formation of an inactive enzyme. Arguing against this model is the existence of "flat revertant" and "abortively" transformed cell lines that continue to carry viral genome but have a normal phenotype and aminosugar transferase activity. However, several studies with virally transformed hybrid cells indicate that the viral genome can be associated with more than one chromosome, possibly in a random fashion (Marin and Littlefield, 1968; Weiss et al., 1968). Then association of viral genome with one particular host chromosome would be required for repression of aminosugar transferase and could explain the non-obligatory nature of the altered ganglioside metabolism.

It is also evident that cellular factors are involved in the expression of the transformed phenotype. The existence of "flat" variants (Pollack et al., 1968) as well as studies on polyoma-transformed hamster cells (Rabinowitz and Sachs, 1969; Hitotsumachi et al., 1971) demonstrate the ability of the host cell to modulate the expression of functions associated with transformation. Little is known about normal cellular regulation of glycolipid synthesis. Based on studies with synchronized populations of mouse cells, there is evidence that glycolipid synthesis occurs only during a short period prior to and during mitosis (Bosmann and Winston, 1970), and is repressed during most of the cell cycle. Since the established mouse cell lines used in our studies have division times of 18–24 h, a comparison of higher ganglioside content and aminosugar transferase activity in the various cell lines suggests that there is an excess of aminosugar transferase activity to account for all of the higher gangliosides (G_{M2}, G_{M1} and GD_{1a}) synthesized during one cell division (cf. Fishman, Brady and Mora, 1972a). Thus, this key enzyme may only express activity during a portion of the cell cycle and be repressed during the remainder of the cell cycle. Presence of a transforming virus in the cell may interfere with this normal regulation of ganglioside synthesis and result in continuous repression of the aminosugar transferase. Since the "flat revertant" cells of Pollack are heteroploid (Pollack *et al.*, 1970), the presence of unaffected chromosomes would permit normal regulation of glycolipid biosynthesis. This mechanism would be compatible with the transformation model of Sachs which envisions a balance between expressive and suppressive chromosomes to explain the transformed state of a cell (Rabinowitz and Sachs, 1969; Hitotsumachi *et al.*, 1971).

CONCLUSION

Transformation of mouse cells in culture by oncogenic DNA viruses and, recently, RNA tumor viruses, is associated with altered ganglioside metabolism in these cells. The change is specific: a drastic reduction in the activity of the enzyme hematoside:N-acetylgalactosaminyltransferase. Though the underlying molecular basis of this change is unknown, it is clearly related to viral transformation *per se* and not to abnormal morphology in culture. Although a correlation between transformation in culture and malignancy *in vivo* has not been conclusively established (Koprowski *et al.*, 1966), the possibility of a relationship between the functioning of N-acetylgalactosaminyltransferase and neoplastic processes *in vivo* is being examined.

REFERENCES

Bosmann, H. B. and Winston, R. A. 1970. *J. Cell. Biol. 45*, 23–33.

Brady, R. O. and Mora, P. T. 1970. *Biochim. Biophys. Acta, 218*, 308–319.

Buck, C. A., Glick, M. C. and Warren, L. (1970). *Biochemistry, 9*, 4567–4576.

Burger, M. M. and Goldberg, A. R. (1967). *Proc. Natl. Acad. Sci. U.S.A., 57*, 359–366.

Cumar, F. A., Brady, R. O., Kolodny, E. H., McFarland, V. M. and Mora, P. T. (1970). *Proc. Natl. Acad. Sci., U.S.A., 67*, 757–764.

Den, H., Schultz, A. M., Basu, M. and Roseman, S. (1971). *J. Biol. Chem. 246*, 2721–2723.

Dijong, I., Mora, P. T. and Brady, R. O. (1971). *Biochemistry 10*, 4039–4044.

Dubbs, D. R., Kitt, S., deTorres, R. A. and Anken, M. (1967). *J. Virol. 1*, 968–979.

Dulbecco, R., Hartwell, L. H. and Vogt, M. (1965). *Proc. Natl. Acad. Sci., 53*, 403–410.

Eckhart, W. (1969). *Nature, 224*, 1069–1071.

Fishman, P. H., Brady, R. O. and Mora, P. T. (1971). *J. Amer. Oil Chem. Soc.*, in press.

Fishman, P. H., McFarland, V. W., Mora, P. T. and Brady, R. O. (1972b). *Biochem. Biophys. Res. Commun. 48*, 48–57.

Fishman, P. H., Bassin, R., Mora, P. T. and Brady, R. O. (1972c). Manuscript in preparation.

Grimes, W. J., Sasaki, T. and Robbins, P. W. (1971). *Fed. Proc. 30*, 1185 Abs.

Habel, K. (1967). *Curr. Top. Microbiol. Immunol. 41*, 85–99.

Hakomori, S. and Murakami, W. T. (1968). *Proc. Natl. Acad. Sci. U.S.A., 59*, 254–261.

Hitosumachi, S., Rabinowitz, Z. and Sachs, L. (1971). *Nature 231*, 511–514.

Inbar, M., Rabinowitz, Z. and Sachs, L. (1969). *Int. J. Cancer 4*, 690–696.

Kampine, J. P., Brady, R. O., Yankee, R. A., Kanfer, J. N., Shapiro, D. and Gal, A. E. (1967). *Cancer Res.*, *27*, 1312–1315.

Koprowski, H., Jensen, F., Girardi, A. and Koprowski, I., *Cancer Research 26*, (1966), *26*, 1980–1987.

Marin, G. and Littlefield, J. W. (1968). *J. Virology 2*, 69–77.

Mora, P. T., Brady, R. O., Bradley, R. M. and McFarland, V. W. *Proc. Natl. Acad. Sci.* (U.S.A.), *63*, 1290–1296.

Mora, P. T., Cumar, F. A. and Brady, R. O. (1971). *Virology 46*, 60–72.

Pollack, R. E., Green, H. and Todaro, G. J. (1968). *Proc. Natl. Acad. Sci.*, U.S.A., *60*, 126–133.

Pollack, R. E., Wolman, S. and Vogel, A. (1970). *Nature*, *228*, 938, 967–970.

Rabinowitz, Z. and Sachs, L. (1969). *Virology*, *38*, 336–342.

Robbins, P. W. and MacPherson, I. A. (1971). *Nature*, *229*, 569–570.

Sakiyama, H. and Burge, B. W. (1972). *Biochemistry*, II, 1366–1377.

Sheinin, R., Onodera, K., Yogeeswaran, G. and Murray, R. K. (1971). *The Biology of Oncogenic Viruses*. IInd LePetit Symposium, pp. 274–288.

Smith, H. S., Gelb, L. D. and Martin, M. A. (1972). *Proc. Natl. Acad. Sci.*, U.S.A., *69*, 152–156.

Smith, R. W., Morganroth, J. and Mora, P. T. (1970). *Nature*, *227*, 141–145.

Tevethia, S. S., Katz, M. and Rapp, F. (1965). *Proc. Soc. Exptl. Biol. Med.*, *119*, 896–901.

Weiss, M., Ephrussia, B. and Scaletta, L. (1968). *Proc. Natl. Acad. Sci.*, U.S.A., *59*, 1132–1135.

Wu, H. C., Meezan, E., Black, P. H. and Robbins, P. W. (1969). *Biochemistry*, *8*, 2509–2517.

Yogeeswaran, G., Schimmer, B. P., Sheinen, R. and Murray, R. K. (1972). *Fed. Proc.*, *31*, 438 Abs.

4 Cell Growth Regulation, Cell Selection and the Function of the Membrane Glycolipids

Peter T. Mora
Laboratory of Cell Biology
National Cancer Institute
National Institutes of Health
Bethesda, Maryland

The work in this laboratory on cell membranes and their role in cell proliferation includes biochemical studies on glycolipids and glycoproteins, and also studies on certain immunological properties of the cells. Generally, we employ established mouse cell lines which propagate indefinitely in tissue culture. This provides us for the biochemical studies continuous and homogeneous cell populations, where no further drastic drift in cell properties or selection of cells occurs. We restricted our studies to mouse, because of the availability of several genetically well-defined and studied inbred strains. We employ cells in culture, this allows us to observe temporal and biochemically well-defined processes which result in cell transformation in growth potential, such as cell transformation by tumorigenic viruses. Choosing viruses, especially DNA viruses as transforming agents, we are of course motivated by a generally assumed conceptual advantage: we hope that the heritable change which occurs in virally induced transformation may be unravelled in biochemical terms by connecting changes on DNA level with phenotypic changes eventually exhibited on the cell surface. Other transforming agents such as carcinogenic chemicals, radiation, etc., of course are also known to endow certain heritable differences in growth and surface properties to cells. However at least some of the surface properties, for example, antigenic properties of these transformed cells, including the so-called tumor specific transplant antigens, are known to be different even in repeated transformation by the same chemical on the same cell line; not to mention the multitudes of biochemical mechanisms which might be involved and the unspecificity of the reactions which might be postulated even on the DNA level. Consider the randomness of the expected intercalation between the bases of the hydrophobic polycylic carcinogenic chemicals, or the randomness of distribution along the DNA of the radiation induced thymidine dimers, or of other changes in the nucleoside bases.

However, all this biochemical and immunological advantage in using established cell lines in culture and the DNA tumour viruses leads to justifiable questions: To what extent are the observed cell surface biochemical changes such as the glycolipid and related enzymological observations influenced by tissue culture conditions? How general are these changes when we consider, for example, various cells from different organs *in vivo*? How much influence is there of cell selection and drift which occurs in culture, when establishing and cultivating cells for prolonged time, and could these lead to some of the observations? Do the phenotypic changes observed such as in the sialic acid containing glycolipids: the gangliosides, only pertain to alteration of cell surface membranes, or as it is more likely, are these changes evident also in other sub-cellular membrane rich components? Are the observed changes in the glycolipids and in the pertinent enzyme activities obligatory in all viral transformations, or are they only conditional or even just unpredictably occasional? Clearly, these changes can be expressed or not expressed as illustrated by our previous work on certain "flat" phenotypically normally growing "revertant" cell lines (Mora and coworkers, 1971). Therefore, the changes are indeed conditional and apparently are under some kind of cellular control. Would the biochemical properties of both types of cell, the biochemically normal cells before and the biochemically abnormal cells after viral transformation, would these biochemical properties resist both the *in vivo* selection by re-implantation into syngeneic host and the selection of cells during reestablishment of cell lines in culture? Are there related observable immunological changes on the cell surface? The following contribution will discuss these questions, and then will bring attention to some of the problems which occur when it is attempted to extend these observations on cell membranes and cell interactions in culture to the control of growth of these cells as tumors when injected into the syngeneic host.

GLYCOLIPIDS AND GLYCOLIPID-SYNTHESIZING ENZYMES IN VARIOUS TISSUE CULTURING CONDITIONS

The major gangliosides of mouse cells, the abbreviations, and the enzymes in the synthesizing pathway are summarized in Fig. 4.1

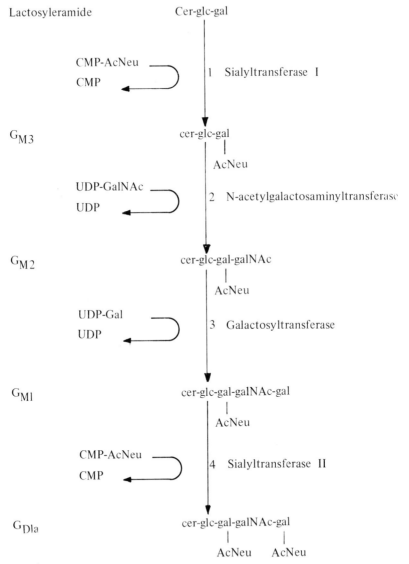

Fig. 4.1 Scheme of ganglioside biosynthesis in mouse cell lines. (From Fishman and coworkers (1972). Abbreviations: G_{M3}, G_{M2}, G_{M1} and G_{D1a} represent the indicated glycolipids in the pathway; AcNeu: N-acetylneuraminic acid.

(see Fishman and coworkers, 1972). The assay of the enzymes, the cell lines and the tissue culture conditions follows our previously published conditions (see Mora and coworkers, 1969; Fishman and coworkers, 1972; Smith and Mora, 1972; cf. also the preceding communication by Brady (1973)).

(1) Freshly cultivated cells obtained from whole mouse embryo, such as AL/N mouse embryo secondary cells were found to possess all the higher gangliosides, including G_{D1a} (Mora and coworkers, 1969). We are currently investigating to what extent the glycolipid pattern and the pertinent enzyme activities would be altered by the following: the method of cell dispersal for tissue plating (by mincing of embryos or tissues, or by trypsinizing); the rapid drifts in cell populations between the "primary" cells and those which are replated the form secondary, tertiary etc. cultures; the known loss of a very large percentage of cells which occurs through the so-called "crisis" period in culture before the cells become established lines. Also what would be the difference when starting cultures from a whole embryo, or from a selected tissue, such as kidney or skin? All of the above parameters affect, at least quantitatively, the distribution of gangliosides and the activity of glycosyltransferases. The viral induced alteration on established cell lines obviously are superimposed changes.

Whole embryos, organs or tumours, etc. can be assayed for gangliosides and transferases, albeit with low accuracy and reproducibility, when expressing for example specific activity of an enzyme. We observed that in these tissues the enzyme activities are generally much lower than in the cultured cell lines obtained from these tissues. The above mentioned cell selection could be partially the reason. Also, the metabolism of a rapidly growing cell in culture reinforced with serum, with amino acids, large amounts of glutamine, plenty of nutrients, carbohydrates, etc. is obviously quite different from the metabolism of most of the cells in tissues *in vivo*. Thus, all of the factors mentioned above must be taken into account when considering our observations on established cells in culture, and the interpretation must be appropriately limited. Much further work is indicated in these lines.

(2) The established cell lines are being generally cultivated in this laboratory under the conditions which these lines were established and then became best adjusted to grow. For example, the Swiss and Balb mouse cell lines are cultivated in Dulbecco-Vogt medium, while the AL/N cell lines are in Eagle's medium which is not quite as enriched in amino acids and vitamins as the Dulbecco-Vogt medium. Frequently we grow cells in open Falcon petri dishes under $NaHCO_3$ buffer in incubators which have moist saturated

air which contains 5% CO_2. Under such conditions there is considerable fluctuation of pH in the medium, especially as cell growth progresses and releases acid metabolites, and this can lead to deviations from the optimal pH for cell growth (cf. Ceccarini and Eagle, 1971). Varying these conditions, however, did not significantly alter in our experience the ganglioside contents, or the glycolipid transferases. Table 4.1 reports activities of the N-acetylgalactosaminyltransferase, the enzyme which was found to be lower in the virally transformed cells, in several cell lines grown under various conditions.

The effect of the serum added to the medium is more important on the ganglioside biosynthesis in cells (Fig. 4.1). In most of our tissue culture medium 10% foetal calf serum was employed to standardize our experiments. At lower serum concentrations, we observed generally lower transferase activities. We have not undertaken yet a thorough investigation of the effects of serum concentration, of serum factors, etc., on specific transferases, although we have observed differences in ganglioside content of certain cells which might be in part caused by the absence or presence of various serum factors. A systematic investigation on well selected cell lines is being planned especially as these might be important in the differential growth and cell metabolism when comparing normal and virally transformed cells (see Holley and Kiernan, 1971).

(3) Very prolonged cultivation of cells, such as over 200 transfers over a two year period, in medium containing 10% foetal calf serum, can lead to partial loss of the higher gangliosides, including the disialotetrasaccharide G_{D1a}, and the reduction or loss in certain transferase activities, somewhat reminiscent of the virally induced changes. However, this was observed only in a few cell lines. For example, in the spontaneously transformed T AL/N cell line after 200 passages in culture there was a decrease in galactosyltransferase and consequently the gangliosides G_{M1} and G_{D1a} were not synthesized (Table 4.2). In the Balb 3T12 cell line after 300 passages there was a loss of N-acetylgalactosaminyltransferase and consequently G_{D1a}. Since tests failed to detect polyoma or SV40 T antigen or murine leukemia virus GS1 antigen, it is unlikely that this result is due in the latter case to incidental viral infection by tumorigenic DNA viruses, or by the derepression and transformation by latent C type murine leukemia virus or by its "oncogene". Spontaneous production of the type C viruses, however, was observed in another laboratory after lengthy cultivation of the Balb 3T12 line (Todaro, 1972). In contrast, in other cell lines, such as in the Swiss 3T3 cells, or the SVS AL/N cells, we did not observe change in the gangliosides or in the glycosyltransferase activities even after more than 300 transfers in culture, and

Table 4.1

ABSENCE OF EFFECT OF CULTURE MEDIUM AND OF BUFFER OR pH IN THE GROWTH MEDIUM ON
HEMATOSIDE N-ACETYLGALACTOSAMINYLTRANSFERASE ACTIVITY IN BALB/c AND IN AL/N CELL LINES
AND IN SV40 TRANSFORMED DERIVATIVE CELL LINES

Cell Line	Growth Medium	Incubation Conditions Buffer	pH	Transferase Activity Δ Radioactivity Incorporated[a] cpm/mg Protein/hr Acceptor: $G_{M3}NG$
Balb 3T3 A31[b] contact inhibited	Dulbecco Vogt	Open, 5% CO_2	$NaHCO_3$	Variable	20,818
	Eagles	Open, 5% CO_2	$NaHCO_3$	Variable	28,104
SV Balb 3T3 50IIA41[b] SV40 transformant	Dulbecco Vogt	Open, 5% CO_2	$NaHCO_3$	Variable	7,265
	Eagles	Open, 5% CO_2	$NaHCO_3$	Variable	10,395
SV Balb 3T3 11A8[b] SV40 transformant	Dulbecco Vogt	Open, 5% CO_2	$NaHCO_3$	Variable	8,627
	Eagles	Open, 5% CO_2	$NaHCO_3$	Variable	5,412
T AL/N passage 50	Eagles	Open, 5% CO_2	$NaHCO_3$	Variable	34,964
	Eagles	Closed	"Tricine"	7.3	32,148
T AL/N passage 150	Eagles	Open, 5% CO_2	$NaHCO_3$	Variable	26,056
	Eagles	Closed	"Tricine"	7.3	21,762
SVS AL/N passage 150	Eagles	Open, 5% CO_2	$NaHCO_3$	Variable	3,135
	Eagles	Closed	"Tricine"	7.3	3,665

[a] Transferase assay as in Cumar and coworkers (1970). The values represent increased synthesis of gangliosides over that of the endogenous synthesis upon adding exogenous acceptor.

[b] Clones from H. S. Smith and coworkers (1971).

[c] Tricine is 50 mM N-tris-hydroxymethyl methylglycine (cf. Ceccarini and Eagle, 1971).

the characteristic difference between the virally transformed and the non-transformed cells were maintained.

(4) To determine whether the slow process of spontaneous changes in prolonged cultivation can be speeded up by cultivating conditions whereby cells are allowed to grow to very crowded conditions, two non-tumorigenic contact inhibited "flat" cell lines, the Balb 3T3 and a clonal derivative Balb 3T3 OA31 line were passed in culture consistently way beyond their confluency. The cells were regularly fed with standard Dulbecco-Vogt medium containing 10% foetal calf serum every two days, and allowed to develop extensive overgrowth. The overgrown cell sheets were replated only every two or three weeks, just before they were ready to come off the plate. After twelve such highly crowded and lengthy passages, which resulted in cells which were indeed growing extensively on top of each other, there was no difference observed in the ganglioside distribution when compared to the original "contact inhibited" cell lines. (Unpublished experiments with R. Brady and L. Danoff.) This is in line with our earlier observation, that spontaneous transformation, selection of cells which grow to higher density in culture and which are tumorigenic *in vivo* do not exhibit the virally induced changes.

In summary mouse cells, whether they are established or freshly cultivated, generally have the full complement of higher gangliosides, including disialotetrasaccharide G_{D1a}, possess the

Table 4.2

GANGLIOSIDE CONTENT GLYCOSYL TRANSFERASE ACTIVITIES AND TESTS FOR VIRUS SPECIFIC ANTIGENS IN MOUSE CELL LINES AFTER EXTENSIVE CULTIVATION

				Viral Specific Antigens		Murine
			Glycosyl Transferase	SV40	PY	Leukemia GS
Cell	Passage #	Gangliosides	Activities[a]	T Antigen		
				CF negative at serum conc.[b]		
T AL/N	40	Full complement	All normal[a]			
	150	Full complement	All normal[a]			
	>200	G_{M1} and G_{D1a} missing	Galactosyl Transferase 4% of normal			
Balb 3T12	<200	Full complement	Normal			
	>300	G_{D1a} missing	Loss of N-acetyl-galactosaminyl-transferase	<1	<1	<1

[a] See Fishman and coworkers, 1972.

[b] Complement fixation tests performed by R. Gilden.

N-acetylgalactosaminyltransferase activity, and maintain stably such components in standard tissue culture for at least 150 transfers ($1\frac{1}{2}$ years), which represents approximately 300 cell generations. This is unaffected by differences in growth conditions in culture, by the density of the cells even when the density of cells was very high. However, very prolonged cultivation of (over two years) certain cells may bring forth decrease in higher gangliosides and the N-acetylgalactosaminyltransferase activity somewhat similar to that which occurs in viral transformation. However, the nature of such change upon extremely prolonged cultivations of certain cells is unknown, but it does not appear to be necessarily associated with spontaneous viral infection (Table 4.2). There are other paired cell lines, where the ganglioside patterns and enzyme activities are unchanged after over three years of continuous cultivation, maintaining the specific differences endowed by the viral transformation.

GANGLIOSIDE AND GLYCOSYLTRANSFERASE CHANGES

(1) Our original investigation asked to what extent cultivated cells growing on a substratum, when they are harvested by various methods, leave behind or carry with them the glycolipid components, and to what extent the growth stage of cells at the harvesting may influence the glycolipid composition. Washed cells (to remove serum which possess various interfering enzyme activities) when still attached to the substratum, do not appear to release into complete medium, or into balanced salt solution, the glycolipids or the transferases. Cells when detached by various means such as by treatment with 0.5% or 0.001 M EDTA (see Mora and coworkers, 1969) or other cells which can be detached easily and in their entirety by a gentle swirling movement of the culture dish, showed no difference in their ganglioside content when the different methods of detachment were employed. Also, moderately gentle trypsinization (0.05%, 15 min, or 0.25%, 20 min, at 37°) did not alter the amount of the higher gangliosides or the enzyme activities such as the N-acetylgalactosaminyltransferase activity (see Fishman and coworkers, 1972). In each case similar results were obtained to our routine harvesting conditions, in which the cells were collected by gentle scraping under saline just before confluency when they are in their logarithmic phase of growth. There was no significant divergence in any enzyme activity depending on cell density at harvest, including the post confluent stage, when normal and virus transformed Swiss and AL/N mouse cells were analysed (Fishman and coworkers, 1972).

After harvesting and before transferase assays the cells were subjected to two types of treatment to obtain homogenization: (a) rapid decompression from 800 psi to disrupt the cytoplasm but not the nuclei (Cumar and coworkers, 1970) and (b) alternate freezing and thawing (Fishman and coworkers, 1972). Both of these procedures gave comparable results and by both procedures all of the transferases are recovered in good yield. Thus the harvesting of the cell, gentle trypsinization of the whole cell or the nature of the break-up of the cell does not make significant difference in the composition of the glycolipids or in the measurable enzyme activities in the whole cell homogenate.

Partial separation of subcellular fractions and determination of the activity of the N-acetylgalactosaminyltransferase in such fractions showed that most of the enzyme activities and the noted differences in the enzyme activities between the cells not transformed by viruses and cells transformed by viruses are found in membrane-rich subcellular fractions, such as in fractions enriched in mitochondria and especially in microsomes (Cumar and coworkers, 1970). Since protein biosynthesis occurs mainly at the membrane-rich microsomal fraction, it is not surprising that it is there that the major amount of enzyme is synthesized, and that it is there that the enzyme activity is controlled and the significant decrease was observed in the N-acetylgalactosaminyltransferase activity in the virally transformed cells.

We have obtained by different methods various subcellular membrane rich fractions. "Crude" cell membrane fractions (obtained by hypotonic lysis and centrifugation, see Smith and coworkers, 1970) showed the presence of both the gangliosides and the enzymes. Extensive purification of cell membranes by a method which is presumed to result only in cytoplasmic or cell surface membranes (Wallach and Kamat, 1966) indeed provided only smooth membrane components which were free of any other structures identifiable by electronmicroscopy (Fig. 4.3). However, the method as outlined in the legend to Fig. 4.2 did not yield sufficient material to detect either gangliosides or enzyme activity by our standard assay procedures. Still, there was in this fraction considerable incorporation of the sialic acid precursor, the radioactive N-acetylmannosamine, which was present during the growth of the cells, apparently into glycoproteins. This observation, however, does not exclude the presence of small amounts of gangliosides and the required transferases in cell surface membranes, especially since the relatively poor yield of recovery of label in the purified membrane preparation. Their relative amount to that in the whole cell, however, must indeed be small (cf. Bosman, 1972; Keenan and coworkers, 1972).

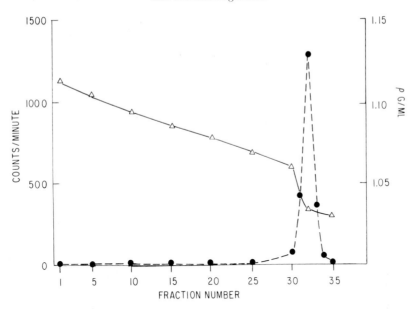

Fig. 4.2 Equilibrium buoyant density distribution in polyglucose gradient of a purified cell surface membrane fraction labelled for sialic acid. An established mouse cell line (designated TAL/N) was grown for two days in tissue culture (Mora and coworkers, 1969) in a medium containing 20% foetal calf serum in the presence of 5 μCi/ml ^{3}H-N-acetylmannosamine. About 0.7% of the input radio-active label was found incorporated into the cells. The washed cells were disrupted by rapid decompression from high pressure (800 psi) (Wallach and Kamat, 1966) in the presence of 0.25 M sucrose and 10^{-3} M Mg in Tris-buffered saline, pH 7.5. The cell homogenate was centrifuged (15 min 15 K in the Sorval centrifuge) to sediment nuclei and mitochondria, and from the supernatant a pellet was obtained by centrifugation (45 min at 40 K in the SW 45 of Spinco Model L). This pellet containing the microsomal fraction and the plasma membrane vesicles was dissolved in 0.5 ml 0.25 M sucrose, and layered on top of a preformed 10–30% (w/v) polyglucose density gradient in a tube of the Spinco SW41 rotor. Centrifugation in the Model L-2 preparative ultracentrifuge was then at 4°C for 3 h at 39,000 rpm, which was shown in earlier experiments to permit a sialic acid labelled pure membrane band (see Fig. 4.3) to attain its equilibrium position (Mora and Luborsky, 1970). Fractions were collected by bottom puncture, and the radioactivity ($\bullet$) and the density (Δ, from refractive index) were determined. The total radioactivity in the peak represented 3.3% of that of the whole cell homogenate.

Most of our observations on the gangliosides, the enzyme participating in ganglioside biosynthesis, and the changes brought forth by viral transformation, obviously pertain mainly to internal cell biochemical structures and controls. Of course, this must bring forth some kind of difference in cell surface membrane composition and function. But for these we did not yet obtain direct evidence on purified cell surface membrane preparations. More work is indicated

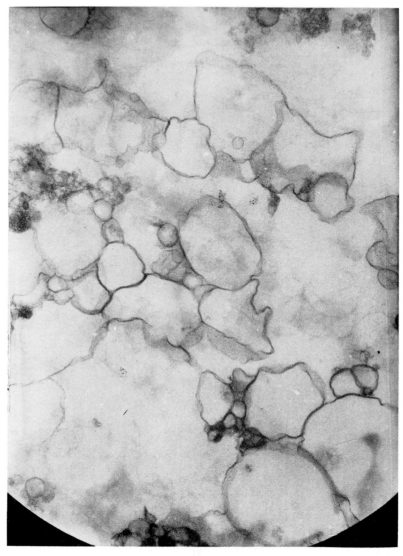

Fig. 4.3 Electron micrograph of a representative area of the pellet from the radioactive fractions 30–34 in Fig. 4.2. Closed vesicles of smooth membranes are the only discernible structures. Some membrane portions show double layer structure, characteristic of cytoplasmic surface membranes. Glutaraldehyde osmium tetroxide fixation. Epon embedded. Section stained with uranyl acetate and lead citrate.

on intact cells and on separated cell membranes, but these experiments must be interpreted especially carefully, for example to avoid interference from "leakage" from a few degrading cells, or from low amount of other cell components present. However, it appears that both the ganglioside composition and the enzyme differences are an inherent and very general property of probably most cellular membranes and may relate to alteration of protein biosynthetic mechanism, associated with predominantly endoplasmic internal cell membranes.

TRANSFORMATION OF CELLS BY SV40 AND POLYOMA VIRUSES

(1) As our previous work on the phenotypically normal flat "revertant" cells show (Mora and coworkers, 1971; see also Brady, this volume, 1973) obviously cells can control the virus induced alteration in higher gangliosides and in the crucial biosynthetic enzymes. In other words, the presence of the virus DNA is not sufficient: cells containing both Py and SV40 DNA but phenotypically growing normally (the "flat revertants" in tissue culture) have high amount of complex glycolipids and N-acetylgalactosaminyltransferase activity, similar to the parent (Swiss 3T3) cell line before the viral transformation. To test how general is such "conjoint control" of cell growth properties in culture and of the glycosyltransferases, we extended our work to other established mouse cell lines. A Balb 3T3 cell line (Clone A31) was transformed with SV40, and after growth in serum factor-depleted medium, various cell lines were selected out with a constant difference in growth properties on substratum, such as in their sensitivity to density dependent, or "contact" inhibition of growth, as measured by the maximum attainable cell density or the "saturation density" on a substratum (H. Smith and coworkers, 1971). Indeed a similar "conjoint" control was found between growth properties in tissue culture and of N-acetylgalactosaminyltransferase activity (see Table 4.3) as previously shown (Mora and coworkers, 1971) for the Swiss 3T3 cells selected out indirectly by resistance to FUDR. The Balb 3T3 cells were selected out by a simple requirement for a serum factor necessary for growth of the virus transformed cells in one direct step. Therefore, different selection procedures direct or indirect, positive or negative, on different cell lines, both can lead to cell lines where there is a close "coupling" between growth properties in tissue culture and in the activity of N-acetylgalactosaminyltransferase.

(2) Numerous cloned and uncloned virus transformed derivative lines from established cloned and uncloned cell lines showed the

Table 4.3
URIDINE DIPHOSPHATE N-ACETYLGALACTOSAMINE:
HEMATOSIDE N-ACETYLGALACTOSAMINYLTRANSFERASE
ACTIVITY IN BALB CELL LINES

Cell line (clone)	Saturation density	Glycolipid acceptor[b] $G_{M3}NAc$ or $G_{M3}NGly$
	cells/cm^2 × 10^{-5}	Δ Radioactivity incorporated[b] (cpm/mg of protein)
Control Balb 3T3 (OA31)	0.5	29,500
SV40 transformed (11-A-8)[a]	4.2	11,250
Flat SV40 transformed (10–74)[a]	0.3	39,000

[a] Clones isolated directly from medium lacking serum growth factor (see Smith, Scher and Todaro, Virology 44: 359, 1971; also Scher and Nelson-Rees, Nature-New Biology 223: 263, 1971).

[b] Average of 4 tissue culture experiments; acceptors N-glycolyl-G_{M3} 0.1 μmole, or N-acetyl-G_{M3} 0.05 μmole.

previously described general difference in glycolipids and N-acetylgalactosaminyltransferase activities after transformation with either SV40 or polyoma. (See Mora and coworkers, 1969, 1971; Brady and Mora, 1970; Dijong and coworkers, 1971.) The clonal cell lines, and the virus transformed derivative clonal lines were obtained from at least three different laboratories. However, two SV40 transformed Swiss 3T3 clones were reported to contain higher gangliosides (Sheinin, 1971).

(3) In an SV40 transformed Balb 3T3 (OA31) clonal derivative cell line (12A3 reclone 3) which does not manifest any phenotypic property of transformation (for example, it does not possess T antigen) and in which the SV40 DNA can only be detected by DNA-DNA C_0t hydrization tests (the so-called "cryptic" transformant, H. Smith and coworkers, 1972), there was also no significant change in the activity of N-acetylgalactosaminyltransferase when compared to the parent Balb 3T3 clone (Table 4.4). Thus when no conventional expression of the viral DNA is manifested such as T antigen or changes in growth properties, there is no decrease in the N-acetyl-galactosaminyltransferase activity.

(4) An SV40 transformed mouse kidney cell line (MKS-BU100) which underwent a complex and lengthy BUdR treatment to develop a line resistant to progressively higher doses of BUdR, after 100 passages and numerous cloning procedures (Dubbs and coworkers, 1967) still possesses high N-acetylgalactosaminyltrans-ferase activity (9.44 nanomoles/mg protein/hr, assayed as in Fishman and coworkers, 1972). However, it must be realized that there is

Table 4.4
PROPERTIES OF TWO BALB CELL LINES

Cell Line	SV40 DNA Equivalents per Diploid Cell[a]	T Antigen[a]	Saturation Density cells/cm^2 × 10^{-5}	N-Acetylgalactos-aminyltransferase Activity[b] Δ cpm/mg protein
Parent Balb 3T3 clone A31	0.39	—	0.5	19,900
"Cryptic" SV40 transformant[a] 12A3 reclone 3	2.7–6.2	—	0.66	17,760

[a] H. S. Smith, L. D. Gelb and M. A. Martin (1972).
[b] Glycolipid acceptor $G_{M3}NAc$, measured as in Cumar and coworkers (1970).

no untreated parent cell line available for comparison and also that the BUdR resistant and cloned cells are highly selected.

(5) In the Swiss 3T3 cell line when transformed by SV40, there is the usual decrease in the N-acetylgalactosaminyltransferase (Table 4.5). However, when such transformed cells were treated repeatedly with FUdR, and then clones selected which showed temperature sensitive modulation of growth in culture (Renger and Basilico, 1972), those transformed cells which grow as transformed cells at 32° but as normal cells at 38° showed no modulation in the N-acetylgalactosaminyltransferase or in two other measured enzyme activities at the two temperatures (Table 4.5); also all the complex

Table 4.5
GLYCOSYLTRANSFERASE ACTIVITIES IN NORMAL AND SV40 TRANSFORMED SWISS 3T3 CELLS AND IN SV40 TRANSFORMED TEMPERATURE SENSITIVE MUTANT SWISS CELLS

Glycosyltransferase	Cell Line Swiss 3T3 Clone ME 32°	38°	Cell Line SV40 Transformed[c] 38°	Temperature Sensitive Mutant Cell H6-15[d] 32°	38°
N-acetylgalactos-aminyltransferase[b]	3.70	3.98	1.67	3.66	3.52
Galactosyltransferase[b]	0.3	0.2		0.5	0.2
Sialyltransferase II[a]		14,150		7,350	5,570

[a] Activity is cpm per mg protein per 2 hours.
[b] Activity is nanomoles per mg protein per 2 hours. The values given represent synthesis of gangliosides using exogenous acceptors as described by Fishman and coworkers (1972).
[c] An SV40 transformed line from Swiss 3T3 clone ME.
[d] H6-15 is a mutant *cell* line, selected out from Swiss 3T3 transformed with SV40 and treated repeatedly with FdU (Renger and Basilico, 1972).

gangliosides found in normal mouse cells were present at both temperatures. It has been pointed out that these cells owe their behaviour to a cellular rather than to a viral alteration, since after fusion of the temperature-sensitive transformed cells with permissive monkey cells, a wild-type SV40 virus was rescued (Renger and Basilico, 1972). Thus in cell mutants which are selected out for differences in growth property in tissue culture at different temperatures, there is no correlation between the enzyme activity and the phenotypic growth property in tissue culture. It may be worth recalling, however, that in the same cell mutant, there was no modulation of another phenotypic property generally accompanying viral transformation, namely independence of gamma-globulin in serum for growth. Therefore these cells are not temperature sensitive mutants in all respects, including the N-acetylgalactosaminyltransferase activity (Table 4.5).

Table 4.6
HEMATOSIDE N-ACETYLGALACTOSAMINYLTRANSFERASE
ACTIVITY IN T AL/N CELL LINE AND IN SV40 TRANSFORMED
DERIVATIVE LINES WHICH WERE FROM TUMOURS PASSED
THROUGH THE SYNGENEIC HOST

Cell Line	Δ Radioactivity Incorporated (cpm/mg protein) on Glycolipid Acceptor $G_{M3}NG$		
Expt:	1†	2†	3†
T AL/N	8,370	31,099	17,420
SVS AL/N	1,536	1,619	3,086
SVS AL/N T	0	NT	NT
SVS AL/N T^2	59	329	2,361
SVS AL/N T^3	NT	799	3,897

NT = Not Tested.
† Independent tissue culture experiments.

(6) Experiments are in progress in this laboratory on transformation by SV40 of both newly established AL/N embryo cell lines and also of continuous AL/N cell lines such as the TAL/N line, followed rapidly by cloning out of single cells as soon as detection of T antigen confirms transformation. Then as soon as sufficient amount of randomly selected T antigen-positive and T antigen-negative clonal derivative cells are available to detect (when present) both the presence of persistent T antigen in substantially all of the progeny cells and also the N-acetylgalactosaminyltransferase activity, we will be able to see how much these two phenotypic expressions will parallel each other in a sufficiently high number of randomly

selected clones. These experiments however, still may not answer the question: Would only those very few cells be transformed which have a common, possibly genetic predisposition to transformation and to attain lower enzyme activity in very short period in tissue culture?

Due to the well known low efficiency of transformation of mouse cells in culture by DNA viruses even at high input of infectious virus, this question might be impossible to answer in this system. The RNA viruses, such as the Moloney Sarcoma virus, however, can transform under favorable condition all the cells in culture within about two days. For this reason, we are studying the kinetics of transformation by the Moloney Sarcoma Virus of the Swiss 3T3 cell line, under conditions when all the cells are transformed and preliminary results also indicate a specific decrease in N-acetyl-galactosaminyltransferase, following one or two days later the morphologic expression of transformation (unpublished results with Dr. Bassin).

In summary a fairly general correlation seem to occur in various mouse cell lines in the phenotypic expressions of the viral genome, in affecting growth control in tissue culture and the N-acetylgalactosaminyltransferase activity. These expressions frequently and closely parallel each other especially when a parent cell line is available for careful comparison. In two cell systems, control of both the transferase activity and the growth property was found to be reversibly coupled. In a cell mutant enzyme activity is not modulated by changes in temperature. It is not clear, how general is the correlation between transformed phenotype as manifested by tissue culture growth property changes and the decrease in the N-acetylgalactosaminyltransferase. Transformation experiments by an RNA tumor virus seem to indicate that the specific change in N-acetylgalactosaminyltransferase activity occurs at least in the majority of the transformed cells, but follows morphologic transformation after one or two days.

GLYCOLIPID COMPOSITION AND ENZYME ACTIVITY

Would the glycolipid composition and the enzyme activity resist *in vivo* selection by transplantation, and also *in vitro* selection when establishing from the tumors new continuous lines in tissue culture?

We have chosen two cell lines obtained from the same parent N AL/N cell line: (1) A cell line, the "spontaneously" transformed T AL/N cells which induce tumors but the cells were not transformed by virus and do not show a decrease in the transferase, and

(2) The SV40 virus transformed SVS AL/N cell line. When these lines were transplanted into adult syngeneic AL/N mice (in case of the SVS AL/N cells the first transplantation was into irradiated syngenic mice to reduce the host's immuno-competence and to prevent rapid rejection, see Smith and coworkers, 1970; Smith and Mora, 1972) the tumors which they produce were then biopsied, trypsinized and replated to form new cultures (SVS AL/N T). Such cultures were examined for both ganglioside composition and activity of the N-acetylgalactosaminyltransferase, before and after the tissue culture "crisis" period in which most cells are lost. The crucial glycolipid transferase differences were maintained, even when such tumor transplant procedures were repeated and the re-established cell cultures (SVS AL/N T^2 and SVS AL/N T^3) were analysed (Table 4.6). Thus, in the SVS AL/N derivative cell lines, the low N-acetylgalactosaminyltransferase activity after viral transformation is stably maintained against repeated selection pressures *in vivo* (the SVS AL/N cells appear to carry strong tumor specific transplantation antigens and are easily rejected (cf. Smith and coworkers, 1970; and Smith and Mora, 1972), and also against selection pressures during re-establishment of the tissue cultures *in vitro*. Similarly, the T AL/N tumor-derived cell line T AL/NT maintains similar high N-acetylgalactosaminyltransferase activity and the similar complex ganglioside components as the parent T AL/N line (unpublished results with Drs. Smith and Fishman). Thus the biochemical traits under study, both those of the non-virus transformed cell lines and those of the virus transformed cell lines, are retained in the few surviving progeny cells, against complex and strong selection pressures.

In conclusion, we observed that virally induced change occurs in mouse cell growth in culture and in cell glycolipid biosynthesis, for which the presence of virus genome is almost always necessary (with the exception of the case of very prolonged cultivation). However, viral transformation may not be sufficient: the expression of the phenotypic properties can be repressed by the cells, such as in the flat phenotypic "revertants" or in those cells which contain SV40 DNA but don't express their T antigen. We don't know the significance, the mechanism or the generality of these changes, or whether they are direct or indirect effects of viral genes. However, the enzyme difference in the cells seems to survive complex selection pressures of tumor transplantation and of re-establishment of cell lines in culture.

TUMORIGENICITY

Clearly, the *in vivo* tumorigenicity of the cells, be they virally or spontaneously transformed, is not directly related to changes in the glycolipids or the transferases; they must be dominated by other cellular and host factors. For example, the spontaneously transformed T AL/N and the derivative T AL/N T^2 cell lines are one of the most tumorigenic cell lines, the Balb 3T12 line is also tumorigenic (Table 4.7), and neither of these cells is a virally transformed cell. In fact SVS AL/N cells should not be considered tumorigenic in the immunologically competent syngeneic host, since they are rejected at high cell dose level. However, selection of SVS AL/N cells by repeated transplantation in syngeneic AL/N mice results in higher tumorigenic cell lines SVS AL/N T, T^2, T^3; and also higher tumorigenic Balb SVT2T and T^2 cells were obtained from the Balb SVT2 cells, by passage through immunologically competent Balb mice (Smith and Mora, 1972). Apparent selection against *in vivo* tumor specific transplant antigens (as contrast to no qualitative difference as measured generally by susceptibility of these cells *in vitro* to serum mediated cytotoxicity in the presence of complement) clearly does not alter the ganglioside content or the transferase activity, while there may occur a great increase in tumorigenicity (Table 4.7). In the *in vivo* tumorigenic potential of cells other cellular and host factors may be important, including those

Table 4.7

TUMORIGENICITY, GROWTH PROPERTIES IN CULTURE, ANTIGENICITY AND GANGLIOSIDE CONTENT OF CELL LINES

Cell Line		Tumorigenic Dose TD_{50}a		"Saturation Density"b	Virus Specific Antigen		Ganglioside
		Host Normal	Irradiated 300	cells/cm^2 × 10^5	T	"TSA"c	Complement
N AL/N		$>10^7$		1.0	−	−	Complete
T AL/N	< 150 passage	10^4	$<10^3$	1.7	−	−	Complete
T AL/N T	< 150 passage	$<10^3$	$<10^3$		−	−	Complete
T AL/N T^2	< 150 passage	$\ll 10^3$			−	−	Complete
SVS AL/N	up to 400 passage	$>10^8$	10^6	3.0	+	+	Deficient
SVS AL/N T	< 150 passage	7×10^5			+	+	Deficient
SVS AL/N T^2	< 150 passage	7×10^5		0.9	+	+	Deficient
SVS AL/N T^3	< 150 passage	10^4		1.2	+	+	Deficient
PY AL/N		10^4		2.4	+		Deficient
PY AL/N T					+		Deficient
Balb 3T3		$>10^8$		0.3	−	−	Complete
Balb 3T12	< 150 passage	10^6		4.3	−	−	Complete
Balb SV T2	< 150 passage	$>10^8$	10^6	12.2	+	+	Deficient
Balb SV T2T	< 150 passage	10^5			+	+	Deficient
Balb SV T2T^2	< 150 passage	$\gg 10^4$			+	+	Deficient

a Number of cells which when injected subcutaneously into 8 week old male syngeneic mice produced tumor in approximately 50% of the recipients (Smith and coworkers, 1970).

b Maximum cell density attainable in 10% foetal calf serum containing medium (Mora and coworkers, 1969).

c Tumor specific antigen detected by cytolysis by specific antiserum or by cell mediated cytotoxicity by immune lymphocytes (Smith and coworkers, 1970; also Smith and Mora, 1972.)

which may prevent or allow recognition of cells by sensitized lymphocytes, those which may be involved in an alteration of the postulated serum blocking factors of tumor antigens, and most importantly other probably independent changes of cell surface and of cell growth control mechanism which may lead directly to the invasiveness of cell and to uncontrolled growth. Clearly neither tumorigenicity, nor tumor specific antigen activity as measured qualitatively by susceptibility to cell lysis by specific antiserum, are directly related to changes in gangliosides and in enzyme activity in ganglioside biosynthesis. It could be, however, that cell surface glycolipids have some indirect effect, such as exposing or covering (antigenic) sites on the cell surface. This would be one explanation for the apparent consistency in the "TSA" expression and the deficiency in the ganglioside complement shown in the last two columns of Table 4.7.

The question remains to what extent changes initiated by viruses in growth properties of cells in culture, i.e.—cell to cell growth regulation on supporting substratum in close proximity or the so-called density dependent or contact inhibition of cell division, pertains to the changes in gangliosides and in the biosynthetic enzymes. It can be stated that generally, when such change in mouse cells is due to DNA viral transformation, it parallels a characteristic glycolipid change, and its expression or repression may be controlled conjointly with changes in certain growth properties in culture, with the exception of highly selected mutant cell lines. It should be reemphasized, that non-viral changes can lead also to loss of contact inhibition (cf. the well known increased saturation density of the 3T12 cells). Also, even in viral transformation there is no necessary correlation between "contact inhibition of growth" as measured by saturation density, the glycolipids, and the tumorigenicity: Note the increased tumorigenicity of the SVS AL/NT T^2 and T^3 lines with lower saturation density when compared to the SVS AL/N cells, with essentially no difference in gangliosides or in transferases. Thus, in virally transformed cell lines, there is no necessary correlation between loss of contact inhibition in culture and increased tumorigenicity *in vivo* and in fact opposite correlation can be demonstrated when other factors, such as the amount of putative tumor specific antigens could be also involved in tumorigenicity. This is important to keep in mind since much current work in viral transformation in culture assumes correlability with tumorigenicity *in vivo*.

To summarize, we observed a cell membrane change, both internal and possibly external, reducible to a specific biochemical difference, which appears to be induced quite generally in cell transformation by tumorigenic DNA and possibly by RNA viruses but not in lytic infection (see Mora and coworkers, 1971). In transformation the continued presence, and the transcription (of at least part of) the virus DNA (such as the T antigen), does not necessarily ensure the expression of the biochemical change, i.e. the decrease in the specific glycosyl transferase activity. The control for such biochemical change could operate at a post transcriptional level, or in some way the viral transformation initiating a heritable alteration of the biochemistry of the cells and the membrane attached protein biosynthesizing apparatus, could depend on different sites of insertion of the viral gene, or could depend on functions of other unknown (derepressed) cellular genes. In considering the alternatives it is worth keeping in mind that recent findings show that there appears to be not much detectable difference in the *in vivo* transcription in transformation and in permissive lytic infection by SV40 (Khoury and coworkers, 1972), and also that SV40 genome insertion, albeit temporarily also occurs in lytic infection into the cell genome (Rozenblatt and Winocour, 1972).

It can be safely stated that the constitutive enzymes for glycolipid and glycoprotein synthesis of normal cells must be under a very complex control. They are probably mutually dependent on each other, on precursors, on products, on feedback processes, and they probably also require close proximity in the membrane to function optimally. It also should be kept in mind that the glycolipids and the glycoproteins in the cell surface are very distal to the DNA in the information transfer process. However, since nucleotide sugar transferases are crucial in endowing the cells with the various glycoproteins and glycolipids, and carbohydrate residues of these are present in cell membranes and indeed some of these hydrophilic sugar chains point outward on the cell surface, and since there is ample evidence that these transferase activities vary in cell cycle, in growth, and in transformation as induced by viruses, it is important that the changes which pertain to these transferases and to the products of such transferases should be further and patiently elucidated. Such is the way to understand to what extent and which cell membrane property or cell surface change pertains to the malignant transformation, be that of viral or of some other origin, and to what extent, and how cells communicate with each other, and how this membrane mediated information regulates their growth *vis à vis* each other both in the simplified conditions *in vitro* and also in the more complex but obviously more important conditions *in vivo*.

Acknowledgements

This work is the result of collaborative undertakings with Drs. R. O. Brady, F. A. Cumar, and P. H. Fishman who carried out ganglioside and the transferase enzyme assays, Dr. R. W. Smith who did most of the immunologic and tumorigenic assays and Mrs. V. W. McFarland who carried out the major share of tissue culture and also some of the *in vivo* tumour transplant experiments. Thanks are also due Mr. Lorenzo Waters for the excellent electron-microscopy and to Mrs. Mattie Owens and Miss Leslie Danoff for assistance with tissue culture.

REFERENCES

Bosmann, H. B. (1972). *Biochem. Biophys. Res. Comm. 48*, 523–529.

Brady, R. O. (1973). *This Symposium.*

Brady, R. O. and Mora, P. T. (1970). *Biochim. Biophys. Acta 218*, 308–319.

Ceccarini, C. and Eagle, H. (1971). *Proc. Nat. Acad. Sci. U.S.A. 68*, 229–233.

Cumar, F. A., Brady, R. O., Kolodny, E. H., McFarland, V. W. and Mora, P. T. (1970). *Proc. Nat. Acad. Sci. U.S.A. 67*, 757–764.

Dijong, I., Mora, P. T. and Brady, R. O. (1971). *Biochemistry 10*, 4039–4044.

Dubbs, D. R., Kit, S., deTorres, R. A. and Anken, M. (1967). *Journ. Virology 1*, 968–979.

Fishman, P. H., McFarland, V. W., Mora, P. T. and Brady, R. O. (1972). *Biochem. Biophys. Res. Comm. 48*, 48–57.

Holley, R. W. and Kiernan, J. A. (1971). In *Growth Control in Cell Cultures*, A Ciba Foundation Symposium, pp. 3–15, Eds. G. E. W. Wolstenholme and J. Knight, Edinburgh and London: Churchill Livingston.

Keenan, T. W., Huang, C. M. and Morre, D. (1972). *Biochem. Biophys. Res. Comm. 47*, 1277–1283.

Khoury, G., Byrne, J. C. and Martin, M. (1972). *Proc. Nat. Acad. Sci. 69*, 1925–1928.

Mora, P. T., Brady, R. O., Bradley, R. M. and McFarland, V. W. (1969). *Proc. Nat. Acad. Sci. U.S.A. 63*, 1290–1296.

Mora, P. T., Cumar, F. A. and Brady, R. O. (1971). *Virology 46*, 60–72.

Mora, P. T. and Luborsky, S. W. (1970). *J. Macromol. Sci.-Chem. A4*, 1067–1078.

Renger, H. C., Basilico, C. (1972). *Proc. Nat. Acad. Sci. U.S.A. 69*, 109–114.

Rozenblatt, S. and Winocour, E. (1972). *Virology 50*, 558–566.

Scher, C. D., Nelson-Rees, W. A. (1971). *Nature-New Biology 233*, 263–265.

Sheinin, R. (1971). In *Proceedings of 1st International Conference on Cell Differentiation*, *Nice*, p. 186–190, Eds. R. Harris and D. Viza, Copenhagen, Munksgaard.

Smith, H. S., Gelb, L. D. and Martin, M. A. (1972). *Proc. Nat. Acad. Sci. U.S.A. 69*, 152–156.

Smith, H. S., Scher, D. C. and Todaro, G. (1971). *Virology 44*, 359–370.

Smith, R. W. and Mora, P. T. (1972). *Virology 50*, 233.

Smith, R. W., Morganroth, J. and Mora, P. T. (1970). *Nature 227*, 141–145.

Todaro, G. J. (1972). *Nature-New Biology 240*, 157–160.

Wallach, D. F. H. and Kamat, V. B. (1966). In *Methods of Enzymology*, Vol. 8, p. 164, ed. Colowick, S. P. and Kaplan, N. O., New York, Academic Press.

5 Applications of Heterophile Agglutinins for Detection of Carbohydrate Structures on Biological Particles Other Than Red Cells

G. Uhlenbruck
Medical University Clinic Cologne
Department of Immunobiology

Heterophile agglutinins from plants (seed extracts), invertebrates (haemolymph, snail albumin gland, snail eggs) and vertebrates (fish sera, fish eggs) have been successfully used in elucidating the chemical nature of blood group specific and other carbohydrate structures (PHA receptor) of the red cell membrane. In preceding publications of our laboratory (Prokop, Uhlenbruck and Köhler, 1968; Pardoe and Uhlenbruck, 1970; Prokop et al., 1972), the main results of which will be summarized here, it has been shown, that these antibody-like agglutinins and precipitins represent pseudo- or para-immunological tools, which agglutinate not only human, but also a large variety of animal red cells (normal, protease- and neuraminidase-treated) on account of their anticarbohydrate specificity. The biological role of these pseudo- and paraimmunological reagents, listed in Table 5.1, is still unsettled, although some suggestions have been made by others and us that the fixation property (of bacteria, viruses, spermatozoa, bacteriophages, carbohydrates of vitamin-like character, opsoninic, cell-lytic, cell-toxic, cell-elimination, anti-tumour(?), parasite-immobilizing, transport etc.) plays a role in defence (recognition, discrimination, phagocytoses, encapsulation etc.), protection or even fertilization. In this connection it is interesting to note, that so far no direct immunological or biochemical relationships between these pseudo- or paraimmunological mechanisms and the immune reactions in vertebrates have been established, although certain parallels can be constructed and assumed (Pardoe and Uhlenbruck, 1970).

Table 5.1
SEROLOGICAL TOOLS FOR THE
INVESTIGATION OF GLYCOSUBSTANCES

I.   *Classical Immunological Tools*
     (a) Specific humoral antibodies (natural, immune)
     (b) Delayed hypersensitivity (specific cellular response)

II. *Pseudo-Immunological Tools*
     (a) Agglutinins from invertebrate haemolymph
     (b) Protectins from snails
         (1) Snail albumin gland
         (2) Snail eggs
     (c) Antibody-like substances from fish roe

III. *Para-Immunological Reagents*
     (a) Lectins from plants
     (b) Agglutinins of fungal origin
     (c) Virus agglutinins
     (d) Agglutinins from bacteria

Most of these heterophile reagents are haemagglutinins and
have been detected by this phenomenon in most cases. This is
outlined in Table 5.2, where also the blood group activity in these
sources is demonstrated. In our studies we have always taken care
to include neuraminidase- and protease-treated cells, when characterizing new agglutinins. Whereas the number of red-cell agglutin-
ating substances progressively increases, very little is known about

Table 5.2
AGGLUTININS AND AGGLUTINOGENS IN PLANTS, FUNGI
VIRUSES AND INVERTEBRATES (WITH SPECIAL REGARD TO
BLOOD GROUPS)

Source	Agglutinins	Agglutinogens	Latest Review
Invertebrates (snails, sponges)	Anti-red cell anti-A, B, P, H, I anti-neuraminyl	P, H, A, AI	Prokop *et al.*, 1972 Khalap, Thompson and Gold, 1970, 1971
Fungi	panagglutinins (animal red cells)	P, N, H, A, B	Reimann and Schulze 1972
Bacteria	anti-animal red cell	P, A, B, H, N	Old 1972 Springer 1966, 1971
Viruses	anti-neuraminyl	A	Pardoe and Uhlenbruck 1970
Plants (Mushrooms)	Anti-A, B, H, P panagglutinins	A, B, H, P	Reimann and Schulze 1972

agglutination reactions with other particles. Therefore we have focussed our experiments more on other blood cells, spermatozoa, subcellular particles of different origin, microorganisms and especially cancer cells.

The aim of this report is, to summarize the essential results of our investigations, to demonstrate some typical examples and to propose general biological principles:

(1) Human lymphocytes as well as platelets, irrespective of blood group, react well before and after treatment with neuraminidase with the agglutinin from the albumin gland of *Helix pomatia*. This is known to detect terminal GalNAc ($=$N-Acetyl-D-galactosamine), α or β linked (Fig. 5.1). The same holds for most animal cells

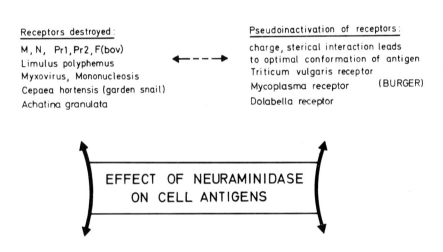

Receptors destroyed:

M, N, Pr1,Pr2, F(bov)
Limulus polyphemus
Myxovirus, Mononucleosis
Cepaea hortensis (garden snail)
Achatina granulata

Pseudoinactivation of receptors:

charge, sterical interaction leads
to optimal conformation of antigen
Triticum vulgaris receptor
Mycoplasma receptor   (BURGER)
Dolabella receptor

EFFECT OF NEURAMINIDASE
ON CELL ANTIGENS

Receptors enhanced or de novo

(Friedenreich antigens)

a.) with terminal D-galactose:

Vicia graminea
Arachis hypogoea
Ricinus communis receptor

b.) with terminal N-acetyl-D-
          galactosamine:

Helix pomatia receptor

„Pseudo - Friedenreich antigens":

agglutination de novo or enhanced
(CURRIE et al.), because

a.) reduction of negative charge
    (repulsion of agglutinin re-
                    duced)

b.) removal of sterical hindering
    groups (HL - A)
            (tumor antigens)

Fig. 5.1   Effect of Neuramidase on all antigens.

of this sort, except for those (for instance pig[HEL]), which carry already an A like receptor a priori. These findings indicate, that GalNAc is located subterminal to the neuraminic acid on the outer membrane of these cells (Uhlenbruck, Vaith and Schumacher, 1972; Heinrich *et al.*, 1972).

(2) Similar observations have been made (Uhlenbruck and Herrmann, 1972; Weis, Uhlenbruck and Schmid, 1972) with spermatozoa of human (group O and B) and animal (bovine) origin. A strong agglutination reaction occurs with the agglutinin from Helix pomatia after these cells have been incubated with neuraminidase. In this connection, we found also, that certain eggs from parasites (nematodes) and fishes are capable of absorbing these para- and pseudo-immunological reagents like an "immunadsorbent", from which these can also be specifically eluted or detected by the addition of the respective red cells, which cause a characteristic mixed cell clumping picture These findings, especially the reaction of sperm cells with so many heterophile agglutinins (see Table 5.3), may have practical importance with respect to

Table 5.3
AGGLUTINATION OF SPERMATOZOA BY HETEROPHILE AGGLUTININS FROM PLANTS AND SNAILS

(Uhlenbruck and Herrmann, 1972)

| | | Spermatozoa | | |
Agglutinin	Origin	Normal	Pronase-treated	RDE-treated
Helix pomatia	bovine*	Ø	Ø	+ + +
	human O†	Ø	Ø	+ + +
	A	+ + +	+ + +	+ + +
Evonymus europaeus	A	+ + +	+ + +	Ø
	O	+	+ + +	Ø
Ricinus communis	O, A	+ + +	+ + +	+ + +
Phaseolus vulgaris	O, A	+ + +	+ + +	+ + +

* Similar results are obtained with lymphocytes and platelets of different origin.

† Strongly after coating with A substance reversible: A active glycoprotein irreversible: A active glycolipid.

sperm-immobilization or egg-coating (possibly preventing the approach of the spermatozoa) and may help to elucidate the question how sperm cells "recognize" the respective ova. As a typical example for heterophile egg cell receptors, the reaction of Dicrocelium dendriticum eggs with Ricinus communis and wax bean agglutinin can be mentioned here (Frank and Uhlenbruck, 1972). Similar investigations using lectins have been performed on protozoa and metazoa by Gerisch (1970).

(3) The agglutination of most bacteria by these agglutinins is understandable, as their cross-reactivity with anti-blood group antibodies (Table 5.2) is very well known (Springer, 1966). Thus several strains of Salmonellae as well as Streptococcus type C can be well distinguished by agglutination absorption with the agglutinin from *Helix pomatia* (Prokop, Uhlenbruck and Köhler, 1968). This reacts also with spike-free Influenza virus, confirming previous results with other heterophile anti-A like agglutinins (Dolichos biflorus) (Klenk, 1971; Klenk, Rott and Becht, 1972). It has been found recently, that like bacteria, fungi also contain agglutinins and blood-group active material, for instance P substance (Reimann and Schulze, 1972). Consequently we followed up these studies and investigated agglutination reactions of fungi with agglutinins from snails and plants (Herrmann and Uhlenbruck, 1972). So we could demonstrate in the case of *Candida albicans*, that it reacted strongly with *Ricinus communis*, wax bean agglutinin, *Vicia graminea* (anti-N), *Evonymus europaeus* (anti-H), *Ulex europaeus* (anti-H) and Concanavalin A, suggesting the presence of blood group active material (H substance) and indicating β-linked terminal D-galactosyl groups on the surface membrane of this fungus (see Table 5.4). Analogous to blood group serology, some receptors are serologically demonstrable only after protease (pronase) treatment, for example the receptor for *Arachis hypogoea* ("anti-T") and that for *Robinia pseudoacacia*,

Table 5.4

AGGLUTINATION REACTION OF CANDIDA ALBICANS WITH CERTAIN HETEROPHILE AGGLUTININS

(Herrmann and Uhlenbruck, 1972)

Agglutinin	Specificity	Agglutination	
		normal	protease-treated
Ricinus	D-gal	+ +	+ +
Evonymus europaeus	D-gal, anti-H	+	+ + +
Ulex europaeus	L-fuc(?), anti-H	+ + +	+ + +
wax bean	D-gal	+	+
Arachis hypogoea	D-gal, anti-T	0	+ + +
Vicia graminea	β-D-gal—galNAc	+	+
Robinia pseudoacacia	D-gal(?)	0	+ + +
Concanavalin A	α-glu, α-man, β-fruc, α-arab	+ +	+ + +

Negative: *Helix pomatia, Triticum vulgaris, Solanum tuberosum, Soja hispida, Laburnum alpinum, Lotus tetragonolobus.*

while others mentioned above show an enhanced α reaction. The question remains still open, whether the agglutinin behaves "incompletely" in these cases (i.e. the agglutinin becomes absorbed without agglutination) or whether an uncovering of these receptors takes place.

(4) As subcellular particles, we have investigated (Voigtmann and Uhlenbruck, 1971) nuclei from chicken and pigeon erythrocytes and found strong agglutination, which could by specifically inhibited, by *Ricinus communis* and *Phaseolus vulgaris* lectin, whereas *Soya hispida* and the agglutinin from the snail *Caucasotachea atrolabiata* reacted only after neuraminidase treatment (Table 5.5). It is interesting to note, that the *Helix pomatia* agglutinin does not agglutinate these neuraminidase-treated nuclei. This is in contrast

Table 5.5

AGGLUTINATION REACTIONS OF CHICKEN AND PIGEON ERYTHROCYTE NUCLEI CAUSED BY SOME HETEROPHILE AGGLUTININS

(Voigtmann and Uhlenbruck, 1971)

Agglutinin	Titre normal	Titre RDE-treated	inhibition titre
Ricinus communis	64	512	8 (D-galactose)
Phaseolus vulgaris	128	1024	16,000 (pig erythocyte mucoid)
Soya hispida	0	16	32 (D-galactose)
Caucasotachea atrolabiata	4	32	32 (N-Acetyl-D-galactosamine)

weak reactions: Maclura aurantiaca, Robinia pseudoacacia, Phaseolus lunatus.
negative results: Helix pomatia, Arachis hypogoea, Cepaea nemoralis, Solanum tuberosum.

to the corresponding red-cell plasma membrane, where, like most other body cells, subterminal GalNAc is uncovered by neuraminidase. Lysosomes, mitochondria and microsomes from rat liver cells can be clumped by *Triticum vulgaris*, *Ricinus communis* and Concanavalin A, a process which is also specifically, inhibited by the respective sugars, with which those agglutinins combine (Uhlenbruck and Henning, 1972). Similar observations have been made by others too (Nicolson, Lacorbiere and Delmonte, 1972). Most exciting were our investigations on subcellular organelles of plants, where we obtained specific agglutination of chloroplasts and thylakoids with *Ricinus communis*, *Phaseolus vulgaris* and the "anti-B" from *Salmo trutta*. In the latter case we are even in a position to localize the respective partner in the organelle surface structure, namely digalactosylglycerol, in which the terminal D-galactose is α-linked and can therefore be recognized by "anti-α-galactosyl" (= "anti-human blood group B") reagents (Uhlenbruck and Radunz, 1972). These results have been confirmed by absorption and inhibition studies too (Table 5.6). Also the reactant partner for the Ricinus agglutinin has been enriched as a glycopeptide.

Table 5.6

AGGLUTINATION PATTERN OF PLANT ORGANELLE SYSTEMS
WITH CERTAIN HETEROPHILE AGGLUTININS
(*ANTIRRHINUM MAIUS*)

(Uhlenbruck and Radunz, 1972)

| Agglutinin | *Stromafree chloroplast ultra-sound Thylakoids* | | | |
	lamellar-system	pronase-treated	free (sediment)	membrane fragm. (supernatant)
Ricinus communis	+ + +	+ + +	+ + +	+ + +
Phaseolus vulgaris	−	+ + +	+ + +	+ + +
Arachis hypogoea	−	+	+ +	−
Salmo trutta "anti-B"	+ +	+ + +	+ + +	+ + +
Helix pomatia	−	−	−	−
coating with A substance	+ .+ +			

inactive: *Solanum tuberosum, Soja hispida, Triticum vulgaris.*

(5) The reaction of tumour cells with heterophile antibody-like agglutinins is of special interest. The number of carbohydrate receptors may increase, appear de novo, become serologically available or decrease during malignant transformation, or may become redistributed in a completely new manner during this process (Burger, 1971; Inbar, Ben-Bassat and Sachs, 1972). The reaction of the tumour-cell with characteristic agglutinins from soybean, *Triticum vulgaris* and Concanavalin A has already extensively studied by these and other groups. Our main interest was concerned with the reaction of the *Helix pomatia* agglutinin and certain tumour cells (Prokop and Uhlenbruck, 1969) and the results of these experiments are listed in Table 5.7. Different methods used for the detection of these heterophile receptors on tumour cells are also applicable to other biological particles (Table 5.8). Further advantage is offered by the absorption technique in combination with the immunelectrophoresis, as single bands in the precipitation pattern may disappear or become diminished as in experiments which cannot be done with the heterogenic vertebrate immunoglobulins of antisera. In this context it would be advisable,

Table 5.7

TUMOUR CELLS WITH TERMINAL GalNAc

(Prokop and Uhlenbruck, 1969)

Tumor cell	Terminal Hexosamine	Detection by
Mouse fibrosarcoma virus transformed chemically induced	α, β-GalNAc	Soy bean *Soya hispida* *Helix pomatia*
Detroit cells	β-GalNAc	*Helix pomatia*
Hela cells	β-GalNAc	*Helix pomatia*
Zajdela hepatoma rat	α-GalNAc	*Helix pomatia* *Cepaea nemoralis*
Baby hamster kidney	α, β-GalNAc	*Helix pomatia*

also to investigate subcellular particles of tumour cells with heterophile reagents. As an example, results with normal rat liver cells
are presented in Table 5.9 (Uhlenbruck and Henning, 1972).

From the use of pseudo- and paraimmunological tools in
tumorimmunology, we want to stress some important aspects,
which seem to us of additional value for this field of research:

1. Detection of tumourcell-characteristic alterations in the archi-
   tecture of the membrane surface with respect to quantitative,
   qualitative and topographic (area-topographic, cryt-topographic)

Table 5.8

METHODS FOR THE DETECTION OF HETEROPHILE
RECEPTORS IN BIOLOGICAL PARTICLES
OTHER THAN ERYTHROCYTES

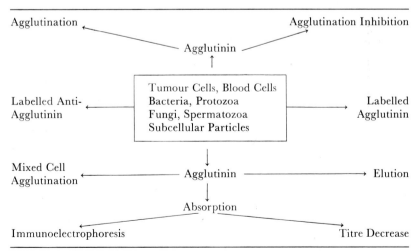

Table 5.9

AGGLUTINATION AND AGGLUTINATION-INHIBITION OF
RAT LIVER CELL PARTICLES BY DIFFERENT
HETEROPHILE AGGLUTININS

(Uhlenbruck and Henning, 1972)

Agglutinin from	Plasma Membrane		Lysosomes*		Mito-chondria	Micro-somes	Inhibitor
	Agglutinin	Inhibition	Agglutinin	Inhibition			
Triticum vulgaris	64	4	2	2	$\emptyset$	$\emptyset$	GlcNAc
Ricinus communis	16	4	32	16	$\emptyset$	$\emptyset$	D-gal
Concanavalin A	128	16	16	4	$\emptyset$ (++)	$\emptyset$ (++)	-D-glc
Phaseolus vulgaris	256	128	32	64	$\emptyset$	$\emptyset$	pig erythrocyte mucoid
Crataegus	++		+		++	++	
Arachis hypogoea	++		+		++	+	
Fomes fomentarius	+				++	+	

Inactive: *Helix pomatia, Solanum tuberosum*   (++) Strongly positive after pronase treatment.
No effect of neuraminidase treatment.   * The possibility of disruption of the lysosomes and probable inside localization of the receptors is fully discussed in the original paper.

changes (Burger, 1971; Inbar, Ben-Bassar and Sachs, 1972; Uhlenbruck and Reifenberg, 1971).

2. Influencing the tumour-cell growth by specific reaction of the membrane receptors with these reagents, especially after labelling with radioactive agglutinins (von Ardenne and Chaplain, 1972; Finck and Vogler, 1972).

3. Applying the red cell linked antigen test with those agglutinins (Sanderson, 1970) to the tumour cell, procedures which may lead also to a better diagnosis of tumours (Prokop and Uhlenbruck, 1969; von Ardenne *et al.*, 1969) and may be valuable for therapy.

4. Treatment of tumour cells with neuraminidase so uncovering of many heterophile receptors de novo or increasing their number, as first suggested by us (Uhlenbruck, 1965) and successfully used now by others (Simmons *et al.*, 1971; Bekesi, St-Arnault and Holland, 1971) as tumour vaccines or immunizing reagents, especially in leukemia.

5. Intensifying attempts to differentiate between T and B lymphocytes with the object of separating T cells and sensitizing them against the respective tumour (Nossal, 1972) or to abolish the "enhancement" phenomenon by influencing B cells.

In summary, it can be expected that the heterophile agglutinins and precipitins will be integrated in the arsenal of specific tools for application in immunobiology and thus increasing our theoretical knowledge.

REFERENCES

Ardenne, M. von, Krüger, W., Prokop, O. and Schnitzler, S. (1969). *Dtsch. Ges. Wes.*, *24*, 588–594.

Ardenne, M. von and Chaplain, R. A. (1972). *Naturwiss.*, *59*, 278.

Bekesi, J. G., St. Arneault, G. and Holland, J. F. (1971). *Cancer Research, 31*, 2130–2132.

Burger, M. M. (1971). In Current Fopies in Cellular Regulation, pp. 135–193, ed. Horecher, B. L. and Studtman, E. R., Academic Press, New York, London.

Finck, W. and Vogler, H. (1972). *Radiobiol.-Radiother.*, *2*, 267–268.

Frank, B. and Uhlenbruck, G. (1972). Unpublished results.

Gerisch, G. (1970). In *Verhandlungsbericht der Deutschen Zoologischen Gesellsch.*, *64.* Tagung. pp. 6–14 Gustav Fischer Verlag.

Heinrich, D., Mueller-Eckhardt, C., Voitmann, R. and Uhlenbruck, G., in II int. Sympol. on Metabolism and Membrane Permeability of Erythrocytes, Thrombocytes and Leukocytes, Vienna 1972, Thieme Stuttgart, in press.

Herrmann, W. P. and Uhlenbruck, G. (1972). *Z. Naturforsch.*, in press.

Inbar, M., Ben-Bassar, H. and Sachs, L. (1972). *Nature*, *236*, 3–5.

Khalap, S., Thompson, T. E. and Gold, E. R. (1970). *Vox sang.*, *18*, 501–526.

Khalap, S., Thompson, T. E. and Gold, E. R. (1971). *Vox sang.*, *20*, 150–173.

Klenk, H. D., personal communication.

Klenk, H. D., Rott, R. and Becht, H. (1972). *Virology*, *47*, 579–591.

Nicholson, G., Lacorbiere, M. and Delmonte, P. (1972). *Exptl. Cell Res.*, *71*, 468–473.

Nossal, G. Y. V. (1972). *Die gelben Hefte, 12*, 1–15.

Old, D. C. (1972). *J. Gen. Microbiol.*, *71*, 149–157.

Pardoe, G. I. and Uhlenbruck, G. (1970). *J. med. Lab. Technol.*, 249–263.

Prokop, O. and Uhlenbruck, G. (1969). *Med. Welt, 20*, 2515–2519.

Prokop, O., Uhlenbruck, G., Ishiyama, I. and Pardoe, G. I. (1970). In "Animal Blood Groups and Biochemical Genetics", Wageningen, Netherlands, in press.

Prokop, O., Uhlenbruck, G., and Köhler, W. (1968). *Vox sang.*, *14*, 321–333.

Reimann, W. and Schulze, M. (1972). *Immun Information, 2*, 28–32.

Simmons, R. L., Rios, A., Ray, P. K. and Lundgren, G. (1971). *J. Nat. Cancer Inst. 47*, 1087–1094.

Sanderson, C. J. (1970). *Immunology, 18*, 352–360.

Springer, G. F. (1966). *Angew. Chem.*, 967–978.

Springer, G. F. (1972). *Progr. Allergy, 15*, 9–77.

Uhlenbruck, G. (1965). Mitteil. *Max Planck Gesellsch.*, *4*, 227–238.

Uhlenbruck, G. and Henning, R. (1972). *Nature*, in press.

Uhlenbruck, G. and Herrmann, W. P. (1972). *Vox sang.*, in press.

Uhlenbruck, G. and Radunz, A., *Z. Naturforsch.* (1972). In press.

Uhlenbruck, G. and Reifenberg, U. (1971). *Med. Klinik, 66*, 1435–1441.

Uhlenbruck, G., Vaith, P. and Schumacher, K. (1972). *Naturwiss.*, *59*, 220–221.

Viogtmann, R. and Uhlenbruck, G. (1971). Unpublished results.

6 *A, B and H Active Blood-group Substances of Human Erythrocyte Membrane*

J. Kościelak, A. Gardas, T. Pacuszka and
A. Piasek
Department of Biochemistry, Institute of Haematology,
Warsaw, Poland

Substances endowed with A, B, H and Lewis blood group specificity occur in the human body as glycoproteins, glycolipids and free oligosaccharides (see Watkins, 1972). Blood group specific glycoproteins are the constituents of secretory fluids as saliva, gastric juice etc. whereas the presence of blood group active oligosaccharides has been so far established in milk (Kobata and Ginsburg, 1969) and urine (Lundblad, 1970). The active glycolipids are the components of cell membranes (Schulman, 1964) and of blood serum (Marcus and Cass, 1969).

Secretory fluids contain by far the largest amounts of blood group substances which are readily obtainable even in gram quantities. On the other hand the contents of the antigens in erythrocyte are low and 30–50 l. of blood are needed in order to isolate few mg of purified substances. Therefore, most information on the structure and biosynthesis of blood group substances comes from the study of the antigens from secretory fluids.

The occurrence of blood group active glycolipids in erythrocyte membranes had been firmly established at the beginning of the preceding decade (Kościelak, 1963; Handa, 1963). Subsequently some of these glycolipids were obtained in a pure state and their qualitative composition was found (Hakomori and Strycharz, 1968; Kościelak, Piasek and Górniak, 1970; Hakomori and Andrews, 1970; Piasek and Kościelak, 1972; Górniak and Kościelak, 1972). The structure of these compounds has not been established although some data on this subject is available (Hakomori, 1970; Kościelak *et al.*, 1970; Iseki, 1970).

Apart from blood group active glycolipids which can be extracted from membrane with organic solvents, erythrocytes contain yet another type of blood group substances which can be extracted with aqueous solvents (Zahler, 1968; Whittemore, Trabold, Reed and Weed, 1969; Gardas and Kościelak, 1971; Yatziw and Flowers, 1971; Kościelak and Gardas, 1971; Liotta, Quintilianni, Buzonetti and Giuliani, 1972). These water extractable blood group A, B and H specific substances seem to be a property of only the erythrocytes of secretors (Gardas and Kościelak, 1971).

The purpose of this paper is to present some new data on the structure and biosynthesis of the blood group active glycolipids and the chemical composition and serological properties of the antigens of the secretors. Evidence is presented that the latter represent a novel hitherto undescribed form of blood group A, B and H substances.

A, B AND H ACTIVE GLYCOLIPIDS

Glycolipids with A, B and H specificity are sphingolipids and occur in erythrocytes of both secretors and non-secretors (Gardas and Kościelak, 1971). They comprise residues of fucose, galactose, glucosamine and of glucose. Glycolipids endowed with A activity contain additionally galactosamine (Hakomori and Strycharz, 1968). Those substances occur in erythrocytes in two and possibly three forms which we designated families (Kościelak, Gardas, Górniak, Pacuszka and Piasek; 1972). Glycolipids of the 1st family comprise a single residue of glucosamine whereas those of the 2nd family have two residues of glucosamine (Table 6.1). Some more

Table 6.1
CHEMICAL COMPOSITION OF ERYTHROCYTE GLYCOLIPIDS,
KOŚCIELAK *et al.*, 1970; KOŚCIELAK AND PIASEK, 1972;
KOŚCIELAK AND GÓRNIAK, 1972

	Specificity		GlucNAc	gluc	gal	fuc	galNAc
	H		1	1	2	1	0
1st family	B		1	1	3	1	0
	A	I	0.94	1	2	0.91	0.94
		II	1	1	2.8	1	1
	H		2	1	3	1	0
2nd family	B		2	1	4	1	0
	A		2.6	1	3	0.9	1.3
3rd family	B		3?				

complex glycolipids, possibly with three glucosamine residues might be also present in erythrocytes (Kościelak et al., 1972) but this result needs confirmation. Glycolipids of the 1st family with B and H blood group specificity were studied by structural methods i.e. by periodate oxidation under conditions used previously (Kościelak et al., 1970) and by methylation. Methylation was carried out according to Adams and Gray (1968). The methylated sugars were analysed by gas chromatographic techniques on columns containing either 3% ethylene adipate or 10% Carbowax 20M coated on GasChrom Q. Methyl ethers of glucosamine were identified employing the technique of Spackman, Stein and Moore (1958) as modified by Donald.

The structure of the B and H active glycolipids of the 1st Family is:

$$\text{Fuc}$$
$$1$$
$$\downarrow$$
$$2$$

B glycolipid
$$\text{gal}(1 \rightarrow 3)\text{gal}(1 \rightarrow 4)\text{glucNAc}(1 \rightarrow 3)\text{gal}(1 \rightarrow 4)\text{gluc-cer}$$

$$\text{Fuc}$$
$$1$$
$$\downarrow$$
$$2$$

H glycolipid $$\text{gal}(1 \rightarrow 4)\text{glucNAc}(1 \rightarrow 3)\text{gal}(1 \rightarrow 4)\text{gluc-cer}$$

It is interesting that only Type II of carbohydrate chain of A, B and H active glycoproteins and oligosaccharides was found in these materials. The glycolipids of the 2nd family comprise a sufficient number of glucosamine residues to form a branched structure. However, the Le^a and Le^b active glycolipids from tumor tissue, as described by Hakomori and Andrews in 1970 comprised a single carbohydrate chain with alternating residues of glucosamine and galactose. Our studies on the structure of the glycolipids of the 2nd Family have not been completed. However, those glycolipids which we have so far analysed all contained only Type II chain.

Another interesting glycolipid, possibly related metabolically to blood group active glycolipids was isolated (Górniak and Kościelak, 1972). This glycolipid is present in erythrocytes at relatively high concentration. It contained sialic acid linked to terminal galactose. The structure of this glycolipid as established by partial acid hydrolysis and methylation was:

$$\text{Sial} \rightarrow \text{gal}(1 \rightarrow 4)\text{glucNAc}(1 \rightarrow 3)\text{gal}(1 \rightarrow 4)\text{gluc-cer}$$

A glycolipid of a similar composition but containing N-glycolylsialic acid was obtained from cattle erythrocytes by Kuhn and Wiegandt

(1964). Wiegandt and Bucking (1970) reported the presence of several sialic acid containing oligosaccharides among degradation products of spleen and erythrocyte gangliosides. The carbohydrate chain of one of these glycolipids had the structure of lacto-N-neotetraose which was identical to that of the presently described glycolipid of erythrocytes. However the presence of a sialo-glycolipid containing lacto-N-tetraose chain (Type I chain) was also reported.

A, B AND H ACTIVE SUBSTANCES WHICH OCCUR IN ERYTHROCYTES OF SECRETORS

First report on the chemical composition of the blood group substances which remain in the aqueous phase after extraction of erythrocyte membranes with n-butanol was that of Kościelak and Gardas (1971) who obtained a substance with B specificity. Subsequently also A and H active preparations were obtained (Gardas and Kościelak, 1972). The isolation procedure involved fractionation of aqueous extracts of membranes on CM cellulose and DEAE-Sephadex columns. The substances were predominantly carbohydrates. However they contained about 7% of amino acids and $1-2\%$ of sphingosine and fatty acids. They were soluble in water in all proportions and insoluble in common organic solvents. They did not migrate on thin layers of silicic acid under conditions employed for the separation of complex glycolipids. Ultracentrifugation of these materials in aqueous solution revealed that they were macromolecules which exhibited sedimentation coefficient $S_{20,w} = 19-20$. However in solvents containing sodium dodecyl–sulphate at 0.2 and 0.5% concentration the sedimentation coefficients of these antigens were lowered to 4 S and about 1.5 S respectively. The latter value of the sedimentation coefficient corresponded molecular weight of 50,000. Sedimenting peaks were symmetrical in both concentrations of sodium dodecyl sulphate employed. Further attempt to disaggregate these substances, employing more concentrated detergent solutions, were unsuccessful. In the latter experiments the molecular weight of erythrocyte antigens was evaluated by gel filtration technique (Gardas and Kościelak, 1972).

The antigens fractionated narrowly with ethanol and the precipitated fractions had the same amino acid composition as the original material. The A, B and H blood group activities of these materials were very high, comparable to those of the most active preparations of the A, B and H specific substances from ovarian cyst, kindly supplied by Prof. W. M. Watkins. This was established in both hemagglutination inhibition and quantitative precipitation tests with a number of sera and reagents of human, animal and plant material. In hemagglutination inhibition tests the material required carrier lipid (Kościelak, 1963) or phospholipids for maximum activity. Le^a, Le^b, P, MN and Rh blood group activities were not found in these antigens. The erythrocyte antigens, unlike blood group active glycoproteins isolated from secretory fluids combine easily with intact erythrocytes with the subsequent change of their specificity. Thus O erythrocytes may be easily changed into B erythrocytes, Bombay erythrocytes may acquire H specificity etc.

The carbohydrate composition of erythrocyte antigens is given in Table 6.2 and the amino acid composition in Table 6.3. A high

Table 6.2
COMPOSITION OF A, B, AND H ACTIVE
BLOOD GROUP SUBSTANCES FROM
ERTHYROCYTES OF SECRETORS.
GARDAS AND KOŚCIELAK 1972

Component	% of substance		
	A	B	H
Galactose	36	42	37
Fucose	13	10	11
Glucose	2	2	2
GlucNH$_2$	29	33	33
GalNH$_2$	4	0	0
Sialic acid	3	1.4	1.8
Acetyl	11	11	11
Amino acids	7.5	7	7

content of N-acetylglucosamine is striking. N-acetylgalactosamine was found only in A substance.

A low content of sphingosine and of fatty acid and the presence of amino acids at constant proportions indicates that these antigens are glycoproteins as previously suggested (Kościelak and Gardas, 1971). However they should be different from typical blood group specific glycoproteins of secretions in that they do not comprise N-acetylgalactosamine in the interior of their carbohydrate chains. This suggests that unlike in secretory blood group substances the linkage of the peptide and carbohydrate portions of the molecule is not of the N-acetylgalactosamine—serine or threonine type. On the other hand, all the analysed preparations contained small amounts of glucose and it is not too unlikely that they might be unusually complex glycolipids. The requirement of these substances for carrier lipid in inhibition tests, their strong tendency towards aggregation in aqueous solution and easy adsorption on erythrocyte membrane suggests that a hydrophobic site (or sites) is present in the

Table 6.3
AMINO ACIDS IN A, B, H ANTIGENS OF THE
ERYTHROCYTES OF SECRETORS. MOLES/100
MOLES OF TOTAL AMINO ACIDS;
GARDAS AND KOŚCIELAK, 1972

	A	B	H
AsP	10.2	14.4	10.1
Thre	7.2	8.7	8.5
Ser	10.2	14.1	10.9
Glu	15.0	11.8	11.7
Pro	5.0	7.7	5.6
Gly	9.3	9.3	9.5
Ala	8.5	6.6	9.2
Val	6.0	6.6	7.0
Ileu	2.4	3.3	4.3
Leu	10.7	5.7	7.6
Tyr	2.3	1.8	3.3
Phe	2.8	2.5	3.3
His	1.8	1.9	2.5
Lys	8.3	5.8	6.9

molecule. If these substances are glycolipids they should be very complex ones with 30–50 residues of carbohydrates per chain, judging from their low glucose and sphingosine content. Such complex glycosphingolipids are unknown in nature and therefore the glycolipid character of the described antigens looks less likely. These antigens might be derived from blood serum. Otherwise it would be difficult to explain how secretor status should be reflected only in these antigens and not in A, B, H specific glycolipids. However these substances are certainly different from serum glycolipids with Le^a and Le^b specificity (Marcus and Cass, 1970).

BIOSYNTHESIS OF MEMBRANE ANTIGENS

Studies on the biosynthesis of membrane A, B, H antigens are few. Blood group active glycolipids may be synthesized in the bone marrow. This tissue is believed to contribute significantly to the level of serum transferase of N-acetylgalactosamine associated with A activity (Schachter, Michaels, Crockston, Tilley and Crookston, 1971). Bone marrow of the rabbit, was also the source of the transferases employed for the enzymic synthesis of ceramide tetrasaccharide, a likely precursor of blood group active glycolipids (Basu and Basu, 1972). However the site of biosynthesis of the glycolipids with Lewis specificity of human plasma (Marcus and Cass, 1970) as well as of the presently described erythrocyte "glycoproteins" may be different.

The purpose of our studies on the biosynthesis of membrane antigens was to establish whether the transferases which synthesize the blood group active determinants of glycoproteins and oligo-saccharides would be active with glycolipid substrates. In the first series of experiments we have shown that alpha galactosyl transferase of milk, which was associated with B specificity could synthesize B character on group O erythrocytes irrespective of the secretor status of blood donors (Pacuszka and Kościelak, 1972). Similar experiments were earlier performed though it had not been possible to distinguish between erythrocytes of secretors and non-secretors. This distinction is however important at present because as shown in this paper two types of blood group substances are present in the erythrocytes of secretors. Finding that erythrocytes of non-secretors are the substrate for the enzyme, has shown that the studied enzymic preparation transferred galactosyl residues to glycolipid precursors. To establish the identity of the enzymes which synthesize B deter-minant in glycoproteins and glycolipids we performed competition studies in which H substance, Le^a substance and 2′ fucosyllactose were allowed to compete for the enzyme molecules with the glycolipid precursors on the surface of group O erythrocytes. As expected H substance and fucosyllactose completely inhibited the synthesis of B determinant on O erythrocytes whereas 10 times higher amounts of Le^a substance were completely ineffective. From these experiments we concluded that Watkins and Morgan (1959) and Cepellini (1959) scheme for the biosynthesis of blood group A, B, H active determinants in secretory glycoproteins is likely to be applicable to blood group active glycolipids.

As to the biosynthetic events which precede the formation of A, B and H active determinants in glycolipids it is conceivable that the likely precursor of blood group active glycolipid is ceramide tetrasaccharide with Lacto-N-neotetraose structure of its saccharide chain. Theoretically this glycolipid might be subsequently branched with galactosyl(1 → 3)N-acetylglucosaminyl unit to form complex glycolipids with type I and type II chains present in the molecule. However, thus far we have not found such glycolipids. Lacto-N-neotetraose structure was also present in the main erythrocyte ganglioside. Therefore it is likely that sialyl transferases which synthesize this glycolipid from ceramide tetrasaccharide competes with transferases responsible for the formation of blood group active glycolipids. It remains to be established whether some rare genetic variants of erythrocytes with only weak expression of A, B, H characters (see Race and Sanger, 1968) do not result from the hyper-activity of this transferase. For example, the high activity of sialyl transferase in ovine submaxillary gland is thought to be responsible for the formation of only short carbohydrate chains in the secreted glycoprotein. Other enzymes which might elongate the chains and specify them with A and H determinants are present in the tissue but they find no substrate to act upon.

CONCLUSIONS

The data presented in this paper indicate that human erythrocytes comprise two types of blood group substances with A, B and H specificity: (a) blood group active glycolipids (b) blood group substances which may be glycoproteins and are encountered only in the membranes of secretors. Blood group A, B and H active glycolipids comprise only Type II carbohydrate chain and presumably have no Le^a or Le^b activity. The absence of Type I carbohydrate chain in these glycolipids would explain why erythrocytes of non-secretors display H but not Le^b specificity. The immediate precursor of A, B and H specific glycolipids is probably a tetrasaccharide with Lacto-N-neotetraose structure of its carbohydrate chain. The latter glycolipid is also a likely precursor of a major sialic acid containing glycolipid of human erythrocytes.

Blood group A, B and H antigens which are present only in erythrocytes of secretors seem to be glycoproteins or unusually complex glycolipids. Absence of Le^a and Le^b activity in these materials indicate that they also comprise only Type II of carbohydrate chain of secretory glycoproteins.

In addition erythrocytes should contain glycolipids with Le^a and Le^b specificity which are derived from serum (Marcus and Cass, 1970). These glycolipids should be wholly responsible for Lewis characters of erythrocyte membrane.

The multiplicity of different chemical species carrying A, B, H and Lewis determinants raises the question of their biological significance. It has been suggested that carbohydrate residues and glycosyl transferases at cell surface play an important role in cell adhesion and contact inhibition phenomena (Roseman, 1970). It would be interesting to test blood group substances in this respect especially as their role in contact inhibition was suggested (Shulman, 1964).

REFERENCES

Adams, J. B. and Gray, G. M. (1968). *Chem. Phys. Lipids 2*, 147.
Basu, M. and Basu, S. (1972). *J. Biol. Chem. 247*, 1489.
Cepellini, R. (1959). *Physiological genetics of human factors*. In Ciba Foundation Symp. on Biochemistry of human genetics p. 242. London, Churchill.
Donald, A. S. R. To be published.
Gardas, A. and Kościelak, J. (1972). *Eur. J. Biochem*. Submitted for publication.

Górniak, H. and Kościelak, J. (1972). *Acta Haemat. Pol. 3*, 175.

Hakomori, S. (1970). *Chem. Phys. Lipids 5*, 96.

Hakomori, S. and Andrews, H. (1970). *Biochim. Biophys. Acta 202*, 225.

Hakomori, S. and Strycharz, G. D. (1968). *Biochem. 7*, 1279.

Handa, S. (1963). *Jap. J. Exp. Med. 33*, 347.

Iseki, S. (1970). In *Blood and Tissue Antigens*, p. 379. Ed. by Aminoff D. New York—London: Academic Press Inc.

Kobata, A. and Ginsburg, V. (1969). *Archives. Biochem. Biophys. 130*, 509.

Kościelak, J. (1963). *Biochim. Biophys. Acta 78*, 213.

Kościelak, J. and Gardas, A. (1971). *Abstr. 7th meeting FEBS, Varna*, p. 220.

Kościelak, J., Piasek, A. and Górniak, H. (1970). In *Blood and Tissue Antigens*, p. 163. Ed. by Aminoff D. New York—London: Academic Press Inc.

Kuhn, R. and Wiegandt, H. (1964). *Z. Naturforsch 19*, 80.

Liotta, J., Quintilianni, M., Buzzonetti, E. and Giuliani, E. (1972). *Vox. Sang. 22*, 171.

Lundblad, A. (1970). In *Blood and Tissue Antigens*, p. 427. Ed. by Aminoff D. New York—London: Academic Press Inc.

Marchesi, V. T. and Andrews, E. P. (1971). *Science 174*, 1247.

Marcus, D. M. and Cass, L. E. (1970). *Science 164*, 553.

Pacuszka, T. and Kościelak, J. (1972). *Eur. J. Biochem.* Submitted for publication.

Piasek, A. and Kościelak, J. (1972). *Acta Haemat. Pol. 3*, 169.

Roseman, S. (1970). *Chem. Physics Lipids. 5*, 270.

Schachter, H., McGuire, E. J. and Roseman, S. (1971). *J. Biol. Chem. 246*, 5321.

Schachter, H., Michaels, M. A., Crokston, M. C., Tilley, Ch. A. and Crookston, J. H. (1971). *Biochem. Biophys. Res. Comm. 45*, 1011.

Shulman, A. E. (1964). *J. Exper. Med. 119*, 503.

Spackman, D. H., Stein, W. H. and Moore, S. (1958). *Analyt. Chem. 30*, 1190.

Watkins, W. M. (1966). *Science, 152*, 172.

Watkins, W. M. and Morgan, W. T. J. (1959). *Vox Sang 4*, 97.

Whittemore, V. B., Trabold, N. C., Reed, C. F. and Weed, R. J. (1969) *17*, 289.

Wiegandt, H. and Bücking, H. W. (1970). *Eur. J. Biochem. 15*, 287.

Yatziw, J. and Flowers, H. M. (1971). *Biochem. Biophys. Res. Commun. 45*, 514.

Zahler, P. (1968). *Vox Sang 15*, 81.

7 Interaction of Adenovirus with Host Cell Membranes

R. C. Hughes and V. Mautner
National Institute for Medical Research, Mill Hill, London

The interaction with the surface of host cells is a critical first step in viral replication. The process requires complementary structures, receptor sites on the host cell surface and the viral component responsible for recognition of these receptor sites. In a number of cases the receptors have been found to be carried by host cell membrane glycoproteins. The best studied examples are the sialoglycoproteins recognized by the myxoviruses. The haemagglutinin molecules in the envelope of these viruses combine with receptor sites containing sialic acid residues present in the carbohydrate chains of membrane glycoproteins of erythrocytes and presumably of permissive host cells. Since terminal residues of sialic acid are commonly found in glycoproteins and plasma membranes are usually rather rich in sialic acid residues it seems unlikely that a single sialoglycoprotein is responsible. In the case of the erythrocyte of course the receptor sites are presumably present on the unique glycoprotein carrying the bulk of the membrane sialic acid. The nature of the receptors for other animal viruses is poorly understood but circumstantial evidence points to membrane glycoproteins being involved in some cases, notably for reoviruses and certain enteroviruses. Reovirus adsorption to red cells was unaffected by treatment of the cells with neuraminidase and sialic acid residues clearly are not involved in the attachment. Presumably the receptor site is located internally to these terminal residues, perhaps as part of the internal sugar sequence (Gelb and Lerner, 1965).

Apart from intrinsic interest in these early events in viral infection, chemical characterization of unique receptor sites for viruses on mammalian cells would provide useful surface membrane markers of the host cells. We describe some observations on the binding of adenovirus type 5, a non-enveloped DNA-containing virus, to plasma membranes isolated from several permissive and non-permissive cells.

MEMBRANE PREPARATION

Fractions enriched in surface membranes were prepared by simple modifications of standard procedures (Warren, Glick and Nass, 1966) as described briefly in Table 7.1. KB cells were grown

Table 7.1
PURIFICATION OF KB PLASMA MEMBRANES

KB cells (approx. 10^8) were washed, swollen in 10 mM Tris-HCl, pH 8, at 2° for 30 min and Dounce homogenized. The intact nuclei and a small (10% or less) proportion of intact cells were removed by centrifugation at 100 × g and 2° for 3 min. The supernatant was fractionated by centrifugation through a discontinuous sucrose gradient (5 ml each of 20, 30 and 50% (w/v) sucrose in 10 mM Tris-HCl, pH 8) at 80,000 × g and 2° overnight. Membranes from the 30–50% interface were harvested and recentrifuged through the same gradient (3 ml of each sucrose solution) and the membrane band at the 30–50% interface was finally purified by centrifugation overnight at 80,000 × g and 2° through a 20–40% linear sucrose gradient as described in Fig. 7.1. Binding of ^{125}I fibre is described in the text.

Fraction	Protein (mg)	Fibre binding (ng/mg protein)	5′-Nucleotidase (Sp. Act.)
Homogenate	48	4.4	1.0
1000 × g pellet	14	3.9	1.1
1000 × g supernatant	31	12	2.7
Membranes, gradient 1	6	19	9.1
Membranes, gradient 2	1.7	38	7.3
Membranes, gradient 3	1.1	87	8.3

at 35° as monolayer cultures to confluence in Leibowitz medium containing tryptose phosphate broth and foetal calf serum. HeLa and L cells were grown at 35° in modified Eagles medium in spinner culture. In a final purification step of KB plasma membranes on a linear sucrose gradient, the 5′-nucleotidase activity profile (Fig. 7.1) coincided exactly with a visible membrane band sedimenting at a density of about 1.15. Different preparations of membranes were enriched about 8 to 15 fold in 5′-nucleotidase (Table 7.1) with overall yields of 10 to 15% of total enzymic activity in the homogenate. It is of course possible that this figure underestimates the yield since it is not established for KB cells that this enzyme is

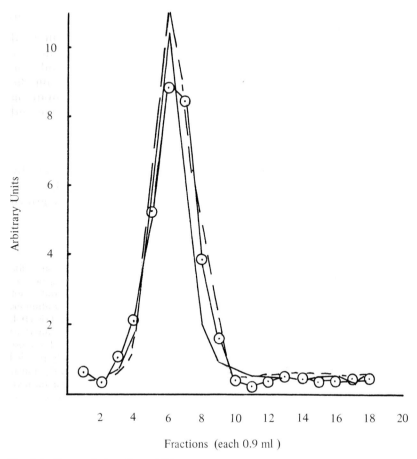

Fig. 7.1 Isopycnic banding of KB plasma membranes. Membranes labelled with [³H]glucosamine (full line) and [³⁵S]methionine (broken line) were centrifuged overnight through a linear 20–40% (w/v) sucrose gradient in a Spinco SW27 rotor at 80,000 × g and 2°. Fractions (0.9 ml) were analysed for radioactivity and for 5′-nucleotidase (0). Direction of sedimentation in this and following figures is to the left.

confined to plasma membranes. Other membrane markers such as cation-activated ATPase and specific binding of adenovirus fibre (discussed later, Table 7.1) were also purified with the plasma membranes. A mitochondrial marker enzyme, succinate dehydrogenase was not detected in purified fractions of KB, HeLa or L cells prepared by the described methods. Plasma membrane material, sectioned and stained with uranyl acetate after fixation with glutaraldehyde followed by osmium, consisted of smooth membranes with negligible contamination by ribosomes and mitochondria as shown by electron microscopy.

Suitably labelled membrane fractions were obtained from cells grown in media containing either [³H] glucosamine or [³⁵S] methionine. Doubly labelled fractions were prepared by mixing appropriately labelled membranes such that equivalent amounts of each isotope were present. On centrifugation of a KB plasma membrane fraction labelled with ³H and ³⁵S through a linear sucrose gradient the two radioactivity profiles followed closely (Fig. 7.1) the sedimentation pattern of 5'-nucleotidase. The incorporation of [³H] glucosamine was shown to be predominantly into glucosamine and sialic acid residues of membrane glycoproteins by paper chromatography, after acid hydrolysis of membrane samples either in 0.1 M HCl at 80° for 1 h to release sialic acid or at 4 N HCl at 100° for 12 h. The relatively poor labelling found for galactosamine suggests that this sugar is a minor component of KB plasma membrane carbohydrates since conversion of the labelled glucosamine precursor to galactosamine as well as sialic acid would have been expected. No conversion of precursor to neutral sugars such as galactose and mannose was detected, and less than 10% of the total incorporated radioactivity was found in the glycolipid fraction.

SUBUNITS

The distribution of protein and glycoprotein subunits in KB plasma membranes was shown by electrophoresis on sodium dodecyl sulphate (SDS) polyacrylamide gels (Fig. 7.2). Gels stained either with amido black or Coomassie blue gave similar but not identical polypeptide patterns. The glycoproteins were confined largely to the higher molecular weight end of the total spectrum (Fig. 7.2) with a group of glycoproteins having molecular weights in the range 80,000 to 280,000. These bands showed up faintly on staining gels with periodate Schiff reagent. In addition the glycolipid fraction of the membranes was detected moving slightly ahead of a bromophenol blue marker. Similar findings were obtained (Fig. 7.3) with HeLa plasma membranes. It remains to be seen if the relatively low mobilities of the glycoproteins in KB and HeLa plasma membranes represent high molecular weights or reflect abnormal behaviour of these membrane glycoproteins on polyacrylamide gels, which tends to overestimate their true molecular weights (Segrest *et al.*, 1971).

The association of intact adenovirus with purified membrane fractions of KB cells was examined (Fig. 7.4) by combining plasma membranes labelled with [³H] glucosamine (spec. act. 1.5×10^5 counts/min/mg protein), with virions prepared by infecting cells in a medium containing [³S] methionine. Labelled [³⁵S] adenovirus

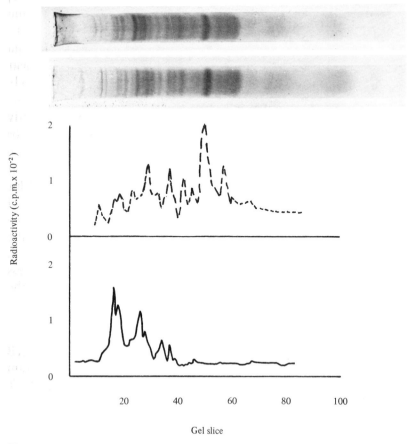

Fig. 7.2 Acrylamide gel electrophoresis of KB plasma membranes. Membranes (200 μg of protein) labelled with [³H]glucosamine and [³⁵S]methionine were solubilized at 90° for 3 min in 1% SDS–1% 2-mercaptoethanol and run on 7.5% acrylamide gels in 0.1% SDS at pH 7. Top gel, stained with Coomassie blue; bottom gel, stained with amido black. A gel was sliced up to the point of migration of a bromophenol blue marker and counted for [³⁵S] radioactivity (top profile) and [³H] radioactivity (bottom profile). Direction of migration in this and following figures is to the right.

(spec. act. 4.1×10^6 counts/min/10^{12} particles) was generously provided by Drs W. C. Russell and H. G. Pereira and particle counting by electron microscopy was performed by Dr N. Wrigley. Sucrose gradient analysis (Fig. 7.4) of the mixtures showed that virus was banded with plasma membranes. The total ^{35}S radioactivity associated with membranes increased with increasing input of virus. At the lowest levels of input the majority of the counts was recovered with the membrane band and at higher levels an increasing proportion of unbound virus was pelleted. Recentrifugation of the membrane-virus complex band through the same

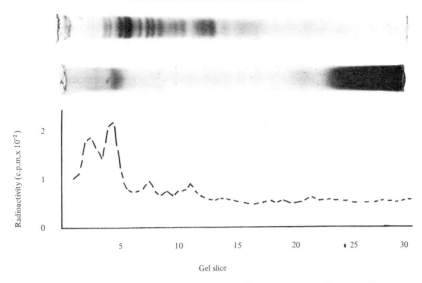

Fig. 7.3 Acrylamide gel electrophoresis of HeLa plasma membranes. The conditions of electrophoresis are given in Fig. 7.2. Top gel, stained with amido black; bottom gel, stained with periodate-Schiff. The fast moving glycolipid fraction stains intensely. Bottom profile, a gel was sliced and glycoproteins labelled with [³H]-glucosamine were located by counting for radioactivity.

sucrose gradient (Fig. 7.5) showed that attachment of virus to membranes was stable and probably not due to non-specific trapping of virus particles. Polyacrylamide gel electrophoretic analysis showed (Fig. 7.5) that virus specific polypeptides labelled with ^{35}S were present in the membrane fraction and these were tentatively identified as known adenovirus components by comparison with published data (Russell and Skehel, 1972). The extent of binding of virus to isolated KB membranes was estimated by the isotope ratios of the complex fractions isolated (Fig. 7.4) by centrifugation, and the known specific activities of membranes and virus. Saturation of binding was not reached in these experiments. At the highest input of virus used $(20 \times 10^9$ particles per $172\,\mu g$ of membrane protein) about 26% of the virions were recovered in bound form. Assuming an average yield of membrane protein from KB cells (Table 7.1) and correcting for losses by the overall yield of 5'-nucleotidase it is calculated that the intact KB cell binds a minimum of $3-6 \times 10^3$ virions. This value agrees approximately with the 10^4 binding sites on intact KB cells for adenovirus type 2 (Philipson *et al.*, 1968).

Solubilization of the membrane-virus complex was readily obtained by treatment with 1% sodium deoxycholate or 1% Triton X100. After incubation the membrane proteins including 5'-nucleotidase were sedimented (Fig. 7.6) through the upper part

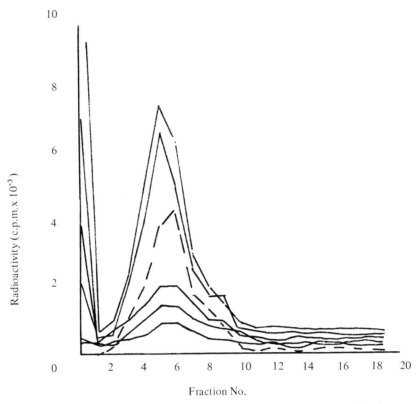

Fraction No.

Fig. 7.4 Binding of intact adenovirus to KB plasma membranes. Membranes (172 μg of protein) labelled with [³H]glucosamine were kept at 37° for 1 h in Leibowitz medium (0.2 ml) with varying amounts of adenovirus labelled with [³⁵S]methionine. The amounts added were 1, 2, 4, 10 and 20 × 10⁹ particles respectively. The mixtures were centrifuged through a linear 10–40% (w/v) sucrose gradient (16 ml) at 2° as described in Fig. 7.1. Fractions (0.9 ml) were analysed for ³H counts (broken line) and ³⁵S counts (full lines). The profile of ³H counts is shown for one representative mixture and followed closely that of 5'-nucleotidase activity (not shown) with a peak of density 1.15. Radioactivity from [³⁵S]-labelled virus was found associated with the membrane band. Unbound virus was pelleted at the bottom of the gradients. Complex fractions sedimenting between fractions 3 and 7 were collected for analysis.

of the gradient. By contrast adenovirus, indicated by ³⁵S radioactivity, was recovered from the pelleted fraction again showing the stability of the virions to mild detergent treatment (Russell, McIntosh and Skehel, 1971). We do not know at present if the specific binding of whole adenovirus to the receptor sites on KB membranes is sensitive to detergent. A small amount of ³H labelled material amounting to less than 5% of the total was sedimented with the pelleted virus. However polyacrylamide gel electrophoresis of this material showed the presence of most of the protein bands

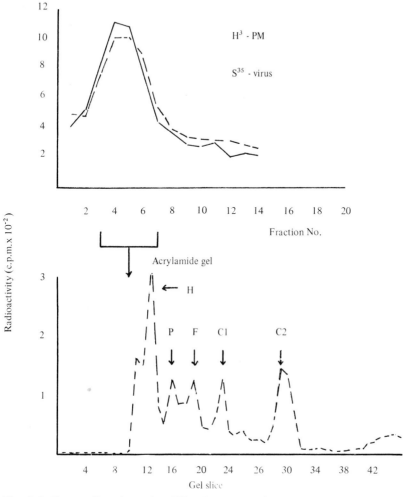

Fig. 7.5 Recentrifugation of a KB plasma membrane-adenovirus complex.
Upper. A complex containing [^{3}H]-labelled membranes and [^{35}S] virus from
Fig. 7.4 was recentifuged through a linear sucrose gradient as previously described.
Solid lines = ^{3}H; broken line = ^{35}S. *Lower.* The complex fraction after recentri-
fugation was examined by electrophoresis on polyacrylamide gels as described in
Fig. 7.2. The gel was sliced and the [^{35}S]polypeptide subunits of the virion were
located by radioactive counting and tentatively identified as hexon (H), penton
base (P), fibre (F) and core proteins (C1 and C2).

detected in the KB plasma membrane (Fig. 7.2) although in very
different relative amounts.

The component of the intact virion responsible for attachment
to KB plasma membranes was identified as the fibre. Membranes
labelled with [^{3}H] glucosamine were incubated with crystalline
structural proteins of adenovirus purified from a lysate of KB cells

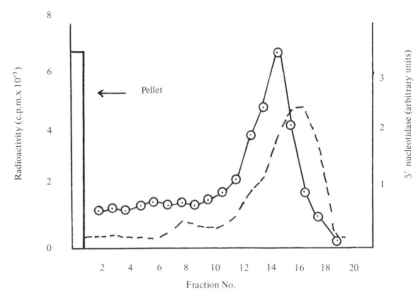

Fig. 7.6   Effect of detergent on the KB membrane-adenovirus complex. Complex isolated as described in the legend to Fig. 7.4 and containing [³H]-labelled membranes (320 μg of protein) and [³⁵S]-labelled adenovirus was kept in 1% sodium deoxycholate (0.5 ml) at 37° for 15 min and then centrifuged through a linear 10–40% (w/v) sucrose gradient (16 ml). The sucrose gradient contained 1% detergent and centrifugation in a Spinco SW27 rotor was at 2° and 80,000 × *g* for 18 h. Fractions (0.7 ml) were collected and analysed for ³⁵S (full line) which was recovered exclusively in the pelleted fraction and for ³H (broken line). The 5′-nucleotidase activity profile is also shown (⊙).

grown in the presence of [³⁵S] methionine (Russell and Skehel, 1972; Pereira *et al.*, 1968; Mautner and Pereira, 1971). After incubation at 37° for 1 h the plasma membranes were banded by centrifugation through a sucrose gradient, as in Fig. 7.4. A portion of the total radioactivity from ³⁵S labelled fibre sedimented with the plasma membranes. The hexon component by contrast was totally recovered at the top of the gradient. In agreement with these results Levine and Ginsburg (1967) and Philipson *et al.* (1968) showed that hexon also did not attach to intact KB cells or inhibit viral infection.

RADIO-IODINATED FIBRES

Radioiodinated fibre was then prepared (specific activity 10⁶ counts/min/μg) and used to examine the interaction with isolated plasma membranes in some detail. The specific binding of ¹²⁵I fibre to isolated KB plasma membranes was a saturable process with increasing concentrations of fibre (Fig. 7.7). The extent of binding was not altered significantly when 1% albumin was included

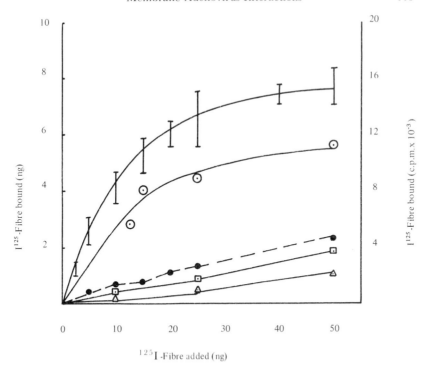

Fig. 7.7 Specific binding of ^{125}I fibre to plasma membranes. Membrane samples (62 μg of protein) were suspended in Leibowitz medium (20 μl) with increasing concentrations of ^{125}I fibre and incubated at 37° for 1 h. Ice cold 10 mM Tris-HCl buffer, pH 7, (0.5 ml) was then added and the suspensions were centrifuged at 2° and 96,000 × g for 1 h. The pellets were washed twice with cold buffer and finally counted for specifically bound ^{125}I fibre. Values obtained in several experiments for specific binding to KB cell membranes are indicated by the *bars*. Representative experiments are shown for Hela cell membranes (⊙), L cell membranes (●), human (⊡) and pig (△) lymphocyte membranes. The lymphocyte membranes were kindly supplied by Dr M. J. Crumpton.

in assay mixtures indicating that the binding was not due to non-specific adsorption. By contrast the presence of non-radioactive adenovirus type 5 fibre in the incubation mixtures drastically reduced the observed binding of ^{125}I fibre (Fig. 7.8), indicating that the iodinated and non-iodinated molecules were competing for the same binding sites. No decrease in binding of ^{125}I fibre was obtained when large amounts (1000 ng) of hexon were incorporated into incubation mixtures containing KB membranes.

A Scatchard plot of the specific binding of ^{125}I fibre to KB membranes (Fig. 7.9) gave a straight line indicating interaction of fibre with a single homogeneous receptor site with no detectable interaction between bound ligand molecules. The average maximal

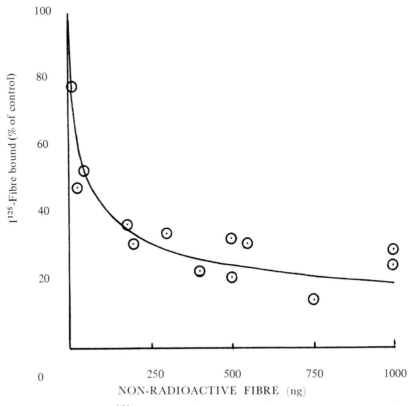

Fig. 7.8 Binding of ^{125}I fibre to KB plasma membranes and displacement by non-iodinated fibre. Membranes (59 μg of protein) were incubated with ^{125}I fibre (25 ng) in Leibowitz solution (50 μl) in the absence and presence of unlabelled fibre. After 1 h at 37° the amounts of fibre bound to the membranes were measured and the percentage of radioactivity recovered in the pellets relative to the control is plotted as a function of cold fibre concentration.

binding capacity of KB plasma membranes is 150 ng of fibre per mg of membrane protein. The average dissociation constant is about 6×10^{-9} M. Again using the approximations previously mentioned it is calculated that there are about $1-2 \times 10^{5}$ binding sites for fibre per KB cell in good agreement with previous results (Philipson *et al.*, 1968).

The rates of formation and dissociation of the fibre-membrane complex are shown in Fig. 7.10. Maximum binding was reached in Leibowitz solution at pH 7 within about 30 min at 37°. Dissociation of the complex in presence of excess fibre is a first order reaction with a half life of 75 min at 37° in 10 mM Tris pH 8. In the absence of excess cold fibre dissociation of the complex was very slow (Fig. 7.10). The association process at 2° showed rapid binding to about 25%

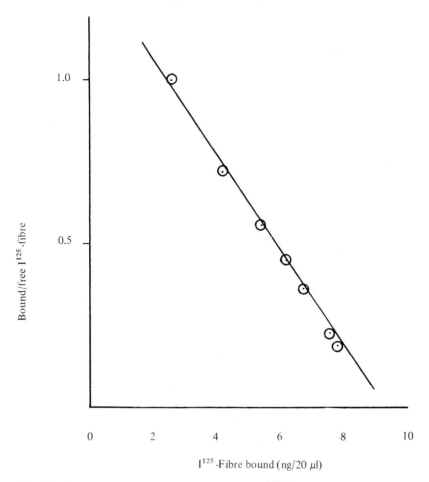

Fig. 7.9 Scatchard plot of the specific binding of ^{125}I fibre to KB plasma membranes. Specific binding is shown in Fig. 7.7.

of maximum within the first hour and a slow rise thereafter. It is possible that a small proportion of high affinity receptor sites are titrated at the low temperature although these were not detected in binding experiments at 37° (Fig. 7.9). Maximum binding of ^{125}I fibre to KB plasma membranes required pH 5 in simple buffers. Comparable binding was obtained in a complex salt solution such as Leibowitz at pH 7. We have not systematically studied the ionic requirements for binding. Maximal binding also depends critically on the final concentrations of reactants and in our experiments incubation mixtures were always below 0.10 ml.

Of the other cell membranes tested none were as efficient in binding ^{125}I fibre as KB cell membranes (Fig. 7.7). Binding to the

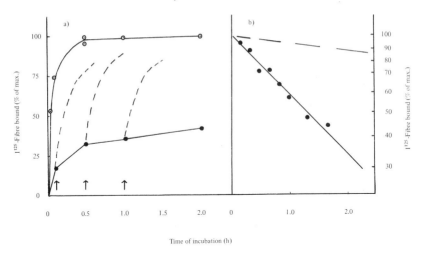

Time of incubation (h)

Fig. 7.10   Rates of binding (a) and dissociation (b) of ^{125}I fibre and KB plasma membranes. (a) Incubation mixtures (20 μl) containing membranes (48 μg of protein) suspended in Leibowitz medium containing ^{125}I fibre (10 ng) were kept either at 2° (●) or 37° (○). At the times indicated samples were diluted with ice cold 10 mM Tris-HCl pH 8 (0.5 ml) and assayed for specific binding. Other incubated mixtures were kept at 2° for 5, 30 or 60 min and then shifted to 37° for either 15 or 30 min before assay of specific binding. The increased binding at 37° is indicated by the broken lines. (b) The rate of dissociation was measured by adding cold fibre (2 μg) to washed membranes (48 μg of protein) previously loaded with ^{125}I fibre by incubation at 37° for 30 min. The samples (50 μl) were incubated at 37° in 10 mM Tris-HCl, pH 8, for varying periods, centrifuged and residual binding of ^{125}I fibre to the membrane pellets was measured. The rate of dissociation of ^{125}I fibre from membranes incubated at 37° in 10 mM Tris-HCl, pH 8, in the absence of cold fibre is shown by the broken line.

plasma membrane fraction of HeLa cells, cells capable of supporting a lytic infection of adenovirus, approximates to that found for KB membranes. Plasma membranes from L cells, human and pig lymphocytes bound fibre poorly. It is already known that L cells are unable to support a lytic infection of adenovirus and the apparent scarcity of specific receptors on their surface may therefore be relevant in this context. We have also found that intact cells do not bind appreciably ^{125}I fibre i.e. less than 5% of ^{125}I fibre bound to intact KB cells and HeLa cells under similar conditions.

The ^{125}I fibre binding activity of KB membranes (Fig. 7.7) was readily destroyed by treatment with trypsin or pronase (Table 7.2). At low concentrations of pronase (1 to 10 μg/ml) however a small stimulation of binding was consistently obtained. Neuraminidase from *Clostridium perfringens* had a somewhat similar effect

Table 7.2
EFFECT OF VARIOUS ENZYMES ON BINDING PROPERTIES OF KB MEMBRANES

Membranes (62 μg protein) were incubated with enzyme at 37° for 30 min and washed by centrifugation. Binding of ^{125}I fibre (10 ng, 20 × 10^3 counts/min) at 37° for 60 min.

Enzyme	Specific binding of ^{125}I fibre	
	c.p.m. × 10^3	% control
None	6.7	100
Trypsin 10 μg per ml	5.1	76
Trypsin 100 μg per ml	1.3	19
Pronase 10 μg per ml	7.6	113
Pronase 100 μg per ml	2.3	34
Neuraminidase 10 μg per ml	8.1	121
Neuraminidase 100 μg per ml	6.7	100

(Table 7.2) at relatively low concentrations (10 μg/ml), giving a slight (10–20%) enhancement of binding. A decreased binding of membranes treated with large amounts of neuraminidase was occasionally found and may be due to proteolytic contamination. Previous conflicting reports have found that treatment of erythrocyte ghosts with neuraminidase did not effect binding of adenovirus type 7 (Neurath et al., 1969) while Kasel et al. (1960) claimed that this binding was destroyed. Clearly however adenovirus type 5 fibre binds efficiently to neuraminidase treated KB plasma membrane, excluding sialic acid residues from participating directly in attachment.

The effect of periodate oxidation on the ability of KB plasma membranes to bind ^{125}I fibre is shown in Fig. 7.11. A very sharp decline in binding activity to about 40% of control was obtained after treatment with rather low concentrations of periodate. No further effect was observed at higher concentrations of periodate. The difficulties of interpretating experiments of this kind are obvious. It is known that degradation of amino acids and proteins may occur (Clamp and Hough, 1965). However the rates of oxidation in these cases are usually slower except possibly for methionine, than are found with carbohydrate units. Therefore the effect on specific ^{125}I binding observed at low concentrations of oxidant (Fig. 7.11) may suggest some role for a membrane component that is particularly sensitive to periodate. It is of interest in this context that terminal sialic acid residues would be very sensitive to mild periodate oxidation because of the acylic glycol side chain at C8 and C9.

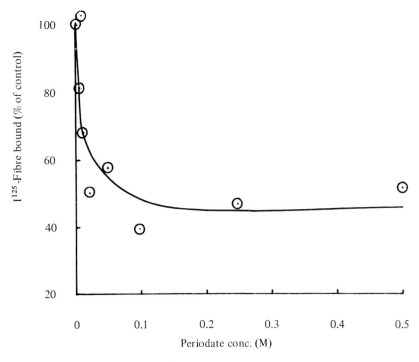

Fig. 7.11 Destruction of adenovirus receptors on KB plasma membranes by periodate. Membrane samples (30 μg of protein) were suspended in solutions (0.1 ml) containing increasing concentrations of sodium metaperiodate in 50 m*M* sodium acetate, pH 5.8. Control tubes contained membranes suspended in buffer alone. After incubation at 37° for 30 min the mixtures were diluted with buffer and membranes were pelleted by centrifugation at 96,000 × *g* for 1 h. The pellets were resuspended in Leibowitz medium (50 μl) containing ^{125}I fibre (25 ng) and the specific binding was measured. Values are expressed as percentages of unoxidized controls.

Indeed Suttajit and Winzler (1971) recently showed that these residues in glycoproteins are degraded to C7 and C8 sugar derivatives by mild periodate oxidation. The modified glycoproteins showed greatly decreased ability to bind myxoviruses. The results just described for adenovirus fibre are clearly not due to a similar mechanism however since sialic acid residues are not involved in the interaction with KB membranes as mentioned earlier.

If it is indeed the case that carbohydrate is involved in the binding of fibre to KB plasma membranes then it is likely that the carbohydrate units would contain sugar residues commonly found in glycoproteins and glycolipids. Many of these residues are known to combine specifically with certain plant lectins and some competition between the binding of lectins and fibre might then be expected if similar sugar residues or sequences were recognized. Since the adenovirus fibre contains no carbohydrate direct interaction between lectins and fibre is unlikely. We examined the effect of Concanavalin A on ^{125}I fibre binding to KB membranes (Fig. 7.12).

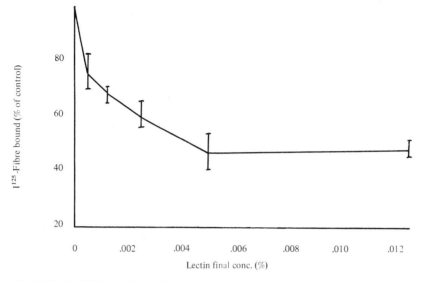

Fig. 7.12 Inhibition of fibre binding to KB plasma membranes by concanavalin A. Membranes (30 μg of protein) were incubated with ^{125}I fibre (25 ng) and increasing concentrations of lectin in Leibowitz medium (0.1 ml) and kept at 37° for 45 min before measurement of binding. Binding is expressed as percentages of controls containing no lectin.

This lectin combines with α-glycosidically linked residues of glucose, mannose or N-acetylglucosamine in glycoproteins and glycolipids. At relatively high concentrations there was a definite inhibition of binding of ^{125}I fibre to KB membranes. The inhibition reached a maximum of 50–60% of control in incubation mixtures where the lectin was in at least 2000 fold excess over the fibre. It is not clear from these data if the lectin and fibre compete for identical binding sites on the KB membrane. The inhibition observed could be due to steric hindrance by the large numbers of concanavalin A molecules bound to the membranes. These could block attachment of fibre either to some adjacent sugar sequence in the same glycoproteins or to other neighbouring sites devoid of carbohydrate but carrying the fibre recognition sites.

CHARACTERIZATION OF SITES

Unequivocal characterization of these sites first requires solubilization and purification of the membrane component responsible. This is obviously an extraordinarily difficult problem since the component makes up at most about 0.015% of the total KB plasma membrane protein as calculable from the binding data in Fig. 7.7. In attempts to solubilize the fibre binding component, KB membranes were treated with deoxycholate or Triton X100. The isolated membranes were readily solubilized by either detergent in the cold or at 37° (Fig. 7.13 and Table 7.3, exp. A). It is interesting

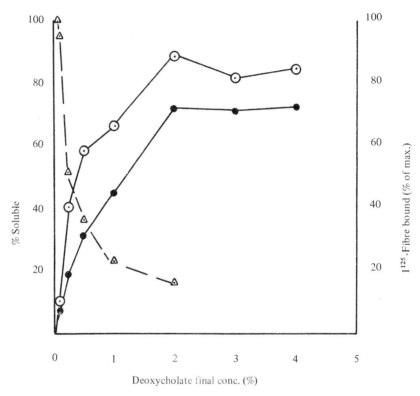

Fig. 7.13 Solubilization of KB membranes by deoxycholate. Membranes (equivalent to 40 μg of protein) labelled with [³H]glucosamine and [³⁵S]methionine were kept at 2° for 45 min in sodium deoxycholate solutions at the stated concentrations and then centrifuged at 96,000 × g for 30 min at 2°. The supernatant fractions were counted for radioactivity and specific binding of ¹²⁵I fibre to the pelleted membrane residues was measured. ³H counts, ⊙; ³⁵S counts ●; ¹²⁵I fibre binding, △.

that in both cases the glycoprotein components were removed selectively from the membrane structure. This is most clearly seen at lower concentrations of detergent. After extraction at 2° for 45 min with 0.5% deoxycholate, 31 and 60% respectively of the total radioactivity associated with the protein and glycoprotein

Table 7.3

SOLUBILIZATION OF KB PLASMA MEMBRANES WITH TRITON X-100

In Exp. A, membranes (840 μg of protein labelled with [^{3}H]glucosamine and [^{35}S]methionine were suspended in 10 mM Tris-HCl, pH 8, incubated for 30 min at 37° with 0.5% Triton X-100 (1.0 ml) and centrifuged at 96,000 × g and 2° for 30 min. The pellet was re-extracted once more at 37° for 30 min and the final membrane residue was suspended in buffer (1.0 ml). Samples (50 μg of protein) of each pellet were assayed for specific binding of ^{125}I fibre. Results are expressed as the amounts of fibre bound to the extracted residues as percentage of that bound to unextracted KB membranes (50 μg). Aliquots (50 μl) of each extract and the final suspended residue were also counted for ^{3}H and ^{35}S radioactivity. In Exp. B, membranes (6 mg of protein) were incubated at 37° for 1 h in Leibowitz medium (1 ml) with ^{125}I fibre (1.25 μg). After washing by centrifugation the membrane pellet contained 0.33 μg fibre. The pellet was suspended in 0.5% Triton X-100–10 mM Tris HCl, pH 8 (2 ml) and kept at 2° for 30 min. The pellet was sedimented at 96,000 × g at 2° for 30 min and re-extracted as before. A third extraction at 2° for 16 h was then carried out. Each extract and the final membrane pellet, suspended in 10 mM Tris HCl, pH 8, was analysed for ^{125}I fibre content. The results are expressed as % of fibre bound to membranes before treatment with Triton X-100.

	% of total		% of control
	^{3}H counts	^{35}S counts	Fibre content
Exp. A.			
Supernatant 1	74	42	—
Pellet 1	—	—	19
Supernatant 2	15	28	—
Pellet 2	11	30	4
Exp. B.			
Supernatant 1	—	—	79
Supernatant 2	—	—	17
Supernatant 3	—	—	<1
Pellet	—	—	<1

components were in soluble form. Since the glycoproteins most probably also contain some methionine residues, the ease of solubilization of these components relative to non-glycosylated membrane proteins is even greater than these values would suggest. Similarly extraction of the major glycoproteins of the KB plasma membrane (Fig. 7.2) was essentially complete after two successive extractions for 30 min each with 0.5% Triton X100 at 2° as shown by SDS-polyacrylamide gel electrophoretic analysis (Fig. 7.14). The membrane residue after extraction contained little glycoprotein material as visualized by periodate-Schiff reagent. Surprisingly a proportion of the membrane glycolipid was not removed under

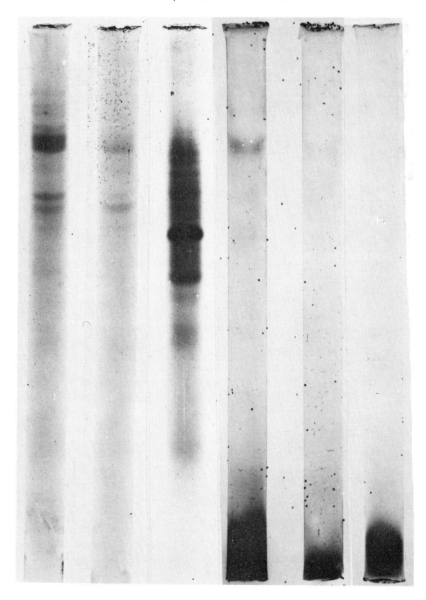

Fig. 7.14 Acrylamide gel electrophoresis of KB plasma membranes treated with Triton X100 at 2°. Membranes (5.6 mg of protein) labelled with [³H]glucosamine and [³⁵S]methionine were extracted twice for 30 min with 0.5% Triton X100 in 10 mM Tris-HCl, pH 8 (2 ml). Each extract and the final membrane residue were examined by electrophoresis in SDS-7.5% acrylamide gels as described in Fig. 7.2. Gels 1–6 from left to right. 1 and 4, soluble proteins in first extract; 2 and 5, soluble proteins in second extract; 3 and 6, proteins in final membrane residue. Gels 1–3 were stained with amido black; gels 4–6 were stained with periodate-Schiff reagent.

these extraction conditions. A notable feature of the extraction was the poor solubilization of the lower molecular weight protein components (Fig. 7.14).

We conclude from these results that the major glycoproteins and several of the non-glycosylated protein subunits with high molecular weights are more loosely associated with the membrane structure. One possibility would be that these represent a class of substances exposed at the exterior surface of the membrane and such a location would of course be logical for molecules carrying viral receptor sites. These conclusions are similar to recent studies with erythrocytes showing that the major glycoprotein and a non-glycosylated protein subunit of relatively high molecular weight (105,000) are externally situated on the membrane (Bretscher, 1971).

Treatment of KB membranes with detergent destroyed the ability of these membranes to bind ^{125}I fibre (Fig. 7.13 and Table 7.3, exp. A). Thus treatment at 2° with 1% deoxycholate for 45 min destroyed greater than 80% of the binding activity. One point to be made about these experiments is the apparent similarity of the curves showing loss of binding ability of detergent treated membranes and the extraction of glycoproteins from the membranes (Fig. 7.13). A similar correlation can be made with the ease of extraction of ^{125}I fibre bound to KB plasma membranes. Thus almost 80% of fibre specifically bound to membranes was removed by 30 min extraction with 0.5% Triton X100 at 2° (Table 7.3, exp. B). A second extraction under the same conditions removed the remaining fibre. The nature of this extracted fibre was examined. Membrane extracts solubilized in Triton X100 and containing the bulk of the iodinated fibre (Table 7.3, exp. B) were eluted through a column of Sepharose 6B with buffer containing the detergent. The fibre was recovered (Fig. 7.15) in a broad peak that almost coincided with the point of elution of stock ^{125}I fibre chromatographed under identical conditions. This suggested that if the interactions between receptor and fibre were preserved during extraction and chromatography then the molecular size of the receptor was small relative to that of the fibre. Otherwise perturbation of the elution of fibre would have been expected. However the elution behaviour of the stock fibre during Sepharose 6B chromatography was not consistent with a known subunit molecular weight of 62,000 and suggested that the fibre was aggregated under these conditions. The sensitivity of this method for estimating the size of molecules attached to the aggregates then becomes rather low. That the fibre-receptor complex was probably isolated intact after Sepharose 6B chromatography however was shown by the following experiment. Fractions (Fig. 7.15) containing ^{125}I fibre were pooled,

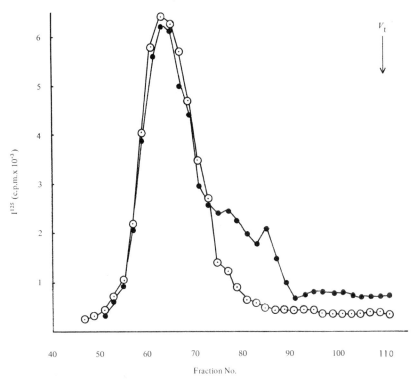

Fig. 7.15 Sepharose 6B chromatography of ^{125}I fibre samples. Free fibre (0.16 μg 1.25 × 10⁵ counts/min) in 1% albumin (0.2 ml) or a Triton X100 extract of KB plasma membranes containing bound fibre (0.16 μg, 1.25 × 10⁵ counts/min), see Table 7.3, exp. B for details, was applied to a column (2 cm × 63 cm) of Sepharose 6B which had been equilibrated with 0.05 mM Tris-HCl, pH 7.2–0.01 M NaCl–0.5% Triton. The column was eluted at room temperature and fractions (1.2 ml) were analysed for radioactivity. The total volume (V_t) was measured by chromatography of [14C]tryptophan (approximately 10⁵ counts/min). Free ^{125}I fibre, ●; ^{125}I fibre extracted from fibre-membrane complex, ⊙.

concentrated by pressure dialysis against 0.05 M Tris HCl pH 7–0.1 M NaCl–0.1% Triton X100 at 2° and tested for specific binding to KB plasma membranes. Membranes (70 μg) were incubated at 37° for 1 h either with the chromatographed ^{125}I fibre (10 ng) or with the ^{125}I fibre (10 ng) extracted from membranes and chromatographed on Sepharose 6B. Measurement of specific binding showed that the membranes bound 26% and 7% respectively of the total ^{125}I fibre added. Clearly the binding ability of ^{125}I fibre present in material extracted from membranes was very much lower than would have been expected. We calculate that at least 75% of the fibre molecules in this preparation presumably are complexed with specific receptors solubilized during the detergent treatment.

In other controls, membranes (70 μg) were incubated with varying amounts of stock ^{125}I fibre (10, 25 and 50 ng) either in 10 mM Tris HCl pH 7 or in this buffer containing 0.1 M NaCl–0.1% Triton X100. Very comparable amounts of specifically bound ^{125}I fibre were obtained in each set of experiments confirming that the concentration of detergent used had little effect on the interaction between fibre and receptors. If therefore these data are taken as evidence that the decreased binding ability of detergent treated membranes is due to the solubilization of receptors, then the correspondence shown between the ease of extraction of these receptors and of the membrane glycoproteins relative to the levels of the membrane proteins may become meaningful. Certainly the selective removal of specific fibre receptors from membranes by mild treatments with detergent makes the task of purifying these receptors less difficult than otherwise.

REFERENCES

Bretscher, M. S. (1971). *J. Mol. biol. 58*, 775–81.

Clamp, J. R. and Hough, L. (1965). *Biochem. J. 94*, 17–24.

Gelb, L. D. and Lerner, A. M. (1965). *Science 147*, 404–5.

Kasel, J. A., Rowe, W. P. and Nemes, J. L. (1960). *Virol. 10*, 388–91.

Levine, A. J. and Ginsburg, H. S. (1967). *J. Virol. 1*, 747–757.

Mautner, V. and Pereira, H. G. (1971). *Nature 230*, 456–7.

Neurath, A. R., Hartzell, R. W. and Rubin, R. A. (1969). *Nature 221*, 1070–1.

Pereira, H., Valentine, R. C. and Russell, W. C. (1968). *Nature 219*, 946–7.

Philipson, L., Lonberg-Holm, K. and Pettersson, V. (1968). *J. Virol. 2*, 1064–75.

Russell, W. C. and Skehel, J. J. (1972). *J. Gen. Virol. 15*, 45–57.

Russell, W. C., McIntosh, K. and Skehel, J. J. (1971). *J. Gen. Virol. 11*, 35–46.

Segrest, J. P., Jackson, R. L., Andrews, E. P. and Marchesi, V. T. (1971). *Biochem. Biophys. Res. Commun. 44*, 390–5.

Suttajit, M. and Winzler, R. J. (1971). *J. biol. Chem. 246*, 3398–3404.

Warren, L., Glick, M. C. and Nass, M. K. (1966). *J. Cell Physiol. 68*, 268–88.

II Biosynthesis of Membrane Components

8 Biochemically-induced Modifications of the Structure of Membrane-bound Oligosaccharides

Paul W. Kent and Peter T. Mora

Glycoprotein Research Unit, Durham University and National Cancer Institute, National Institute of Health, Bethesda, Md., U.S.A.

Accumulating evidence suggests that some oligosaccharides, whether linked either to lipids (as the glycolipids) or to peptide (as in glycoproteins), can contribute to biological recognition processes. This is of particular importance where such oligosaccharide-bearing molecules are located externally on mammalian cell membranes and where they thus confer distinctive informational properties to the cell vis-a-vis its environment. Foremost amongst such situations are the distinctive and well-known blood groups specifications of human red cells and secretions, where the same terminal monosaccharide structures in either glycolipids or glycoproteins provide the determinant groups (as discussed earlier in this volume pp. 95 and by Watkins, 1970).

Here the determinant group is constructed from α-linked L-fucose residues and α-linked D-galactopyranose (in blood group B) or α-linked N-acetyl-D-galactosamine (in blood group A). Where sialic acid residues are present, they would not appear to be concerned with determining specificity. This is in contrast to other examples where sialic acid-bearing glycoproteins and glycolipids are concerned in cell-surface recognition, where their removal by neuraminidase treatment results in considerable alteration of the surface properties (e.g. Currie and Bagshaw, 1968; Winzler, 1970) such as contact inhibition or tumorigenicity. The outstanding work of Ashwell and his colleagues also points to the role of sialic acid residues in α_1-seroglycoprotein in rat blood and ceruloplasmin

(though not in transferrin or certain other glycoproteins) contributing to the "nativeness" of the glycoprotein and to their
persistence in circulation. Removal of even a part of the sialic
acid components results in the rapid withdrawal of the macro-
molecule by absorption into parenchymous cells of the liver.
It is tempting to speculate that such a partly de-sialated structure
has become more "foreign" and therefore susceptible to rejection
(Morell et al., 1971). The further possibility is thus raised that
immunological determinants and other recognition sites may be
concealed by extended glycosylation, and in particular by the
presence of specifically located sialic acid residues (Apfel and
Peters, 1970).

The phenomenon is also observed in the wider context. The
specificity of interaction of a bacteriophage and a bacterial cell
depends on a membrane-bound glycolipid (Losick and Robbins,
1967), genetic modification of which leads to failure of recognition
and emergence of recognition sites for other phage. The invasive
capacity of some viruses, e.g. mumps virus similarly is governed by
the completeness of the virion particle, and in particular by the viral
glycoprotein layer (as discussed by Blough, Gallagher and Weinstein,
p.183).

Recent literature on the structure and role of membrane-
bound glycoproteins has been reviewed recently in detail by
Hughes (1973), Cook (1968) and Roseman (1970).

Biosynthetic Considerations

Though it has to be admitted that in the majority of glycoproteins the function of the carbohydrate is still unknown, in those
cases where its participation in recognition sites is possible, it
becomes of particular interest to elucidate the mechanisms by which
the biosynthesis of the structures are controlled (Kent, 1972).

In principle, distinction can be made between the primary
biosynthetic steps in which assembly of a peptide occurs at polysomal
surfaces under the direct influence of the genome via m-RNA,
according to the commonly accepted mechanism and between
subsequent modification by glycosylation which is not so directly
influenced. Intracellularly-synthesized peptide is transported into
the cisternae in close proximity to the points of attachment to the
ribosomes. The key stage, in which nascent glycoprotein is formed,
occurs on the addition of the first monosaccharide residues to the
peptide on its passage through the membrane or immediately
thereafter (Hallinan, Murty and Grant, 1968; Schenkein and
Uhr, 1970). Current theories implicate the adjacent amino-acid
sequences determining the point of glycosyl attachment to the

peptide (Marshall and Neuberger, 1970; Winterburn and Phelps, 1972). It is a matter for further investigation whether such sequences specify glycosylation points, e.g. threonyl, seryl or asparaginyl or cysteinyl residues with enhanced chemical reactivity or whether they provide sites which serve to localize the necessary sugar transferases in correct stereochemical position to accomplish glycosylation.

Subsequent glycosylations are accomplished by a series of membrane-associated transferases, accomplishing the final oligosaccharide structure according to the kinetics and specificity of each with respect to the sugar nucleotide and to the acceptor. The genetically controlled availability of the transferases must always define ultimately the range of oligosaccharide structures which may be realized. The absence of strictly repeated patterns of monosaccharides in such structures and the existence of microheterogeneity amongst the more complex heterosaccharide attachments gives grounds for the view that the final structures arise through the action of competing transferases in the later stages of glycosylation, especially in association with the Golgi apparatus where this appears generally to occur.

Though comparatively little information exists about the accessibility of such compartmentalized transferases, it is apparent that any selective inhibition of one of the transferases may be expected to give rise to incomplete or altered oligosaccharide structures. A similar result may also be invoked by agents which depress the supply of a particular sugar nucleotide (Winterburn and Phelps, 1971) or which prevent the synthesis of appropriate acceptor molecules (Kent and Mora, 1973).

Effect of Monosaccharide Analogues on Oligosaccharide Biosynthesis

In the majority of studies of glycoprotein biosynthesis by cellular preparations or by particulate enzyme preparations, requisite sugars (or sugar phosphates or sugar nucleotides) are generally employed. In glycoproteins the sugars concerned are usually limited to sialic acids fucose, galactose, N-acetylgalactosamine, N-acetylglucosamine and mannose. In low concentration (e.g. less than 1 mM), D-glucosamine is rapidly incorporated by glycoprotein-synthesizing cells, after phosphorylation and conversion to the UDP-derivative. In high concentrations (e.g. 10 mM or more) noticeable biochemical and cytological changes are observed (Quastel and Cantero, 1952; Molnar and Bekesi, 1972; Bekesi and Winzler, 1969; Richmond, Glasser and Todd, 1968; Bosmann, 1971). Galactosamine and mannosamine exhibited similar but quantitatively lesser effects. The underlying biochemical

cause of these effects whether on specific enzymes and their organiza-
tion or on transport is yet to be established. Sugar analogues, e.g.
2-deoxy-D-glucose in concentrations between 0.1 and 3 mg/ml also
produces disturbances in the synthesis of glycoprotein bound
oligosaccharides. Attachment of radioactive glucosamine mannose,
galactose and fucose to the intracellular carbohydrates of IgG,
myeloma protein in tumour plasma cells is substantially inhibited.
It also appears to inhibit the transfer of the molecules from rough to
smooth membranes (Melchers, 1973).

Present evidence indicates that some fluoromonosaccharides,
i.e. monosaccharides in which OH has been chemically replaced by
F may also restrict synthesis of the oligosaccharide attachments.
(These sugar analogues and other fluorinated metabolites have
been reviewed in detail (C.I.B.A. Symposium, 1972).)

Amino Sugar Analogues

Amongst the various sugar analogues available, particular
attention has been given to N-haloacetylaminosugars, e.g. N-
fluoroacetyl-D-glucosamine (GlcNAcF). This sugar is known to be
stereochemically closely related to N-acetyl-D-glucosamine and to be
capable of occupying the same binding sites in an enzyme, e.g.
lysozyme (Dwek, Kent and Xavier, 1971; Butchard *et al.*, 1972).
In other respects, in contrast to the N-iodoacetyl analogue (GlcNAcI)
(Kent, Ackers and White, 1968) it appears to be relatively unreactive.
The mouse cell lines T AL/N and its SV-40 transformed variant,
SVS-AL/N were cultured in the logarithmic phase of growth for
16 h (i.e. for one cell generation) in the presence of N-fluoroacetylglucosamine (GlcNAcF) and its N-iodoacetyl analogue (GlcNAcI).
Protein biosynthesis was monitored by the incorporation of L-[^{3}H]-
threonine and oligosaccharide synthesis by D-[^{14}C]glucosamine.
The double labelling of macromolecular fractions derived from the
cells (Fig. 8.1) thus enabled distinction to be made in the different
ratios of synthesis of peptide and carbohydrate moieties of the glycoprotein component. In parallel experiments extraction and fractionation of labelled glycolipids by well-established techniques
(e.g. Brady and Mora, 1970) enabled changes in biosynthesis of the
lipid-bound oligosaccharide chains to be detected.

Below 1 mM, both modifiers GlcNAcF and GlcAcI resulted
in increased uptake of ^{14}C-glucosamine by whole cells, with relatively
little change in the ^{3}H-threonine uptake (Fig. 8.2). At higher
concentrations, both caused decrease in the ratio of ^{14}C/^{3}H and in
the case of GlcNAcI, cells were killed. The fluoroanalogue (up to
5 mM) however did not appear to produce toxic symptoms either
in the appearance of cells or in their rate of growth, even after
prolonged culture (6–8 days).

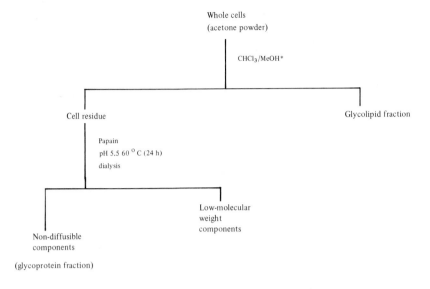

Fig. 8.1   Isolation of oligosaccharide fragments from whole cells.

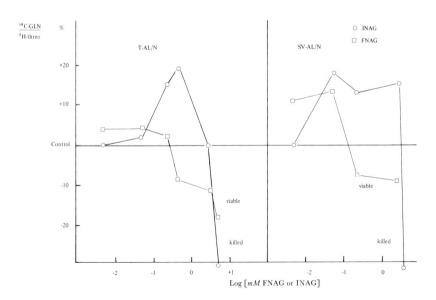

Fig. 8.2   Effect of modifiers on the relative incorporation of ^{14}C-glucosamine (^{14}C-GlN) and ^{3}H-threonine (^{3}H-threo). See Table II. FNAG = N-fluoro-acetylglucosamine (GlcNAcF), INAG = N-iodoacetylglucosamine (GlcNAcI).

Whole cells (TAL/N) grown with ^{14}C-glucosamine for 16 hours had a specific activity of 1.57 Ci/mg cell protein compared with 0.16 Ci/mg cell protein for the cells grown simultaneously in 0.5 mM-GlnAcF. Similarly, the virus-transformed cells had comparable specific activities of 1.73 for controls and 0.49 Ci/mg cell protein for GlcNAcF-grown cells. Growth in the presence of the modifier at this count return did not appear to change markedly the ratio of protein synthesis (Table 8.1). Subsequent experiments also indicated that the transport of amino-sugar metabolite into the cell was not impaired.

Table 8.1

LABELLING OF T AL/N AND SVS AL/N CELLS BY D[1-^{14}C]GLUCOSAMINE, L[G-^{3}H]-THREONINE AND [C-^{3}H$_3$] THYMIDINE. (% OF INPUT COUNTS)

	Cell protein mg/per plate	^{3}H-thymidine incorporation		^{14}C-glucosamine incorporation		^{3}H-threonine incorporation		^{14}C-glucosamine / ^{3}H-threonine
		Total μCi	% (of input)	Total μCi	Kcpm/mg protein	Total	Kcpm/mg protein	(Ratio of specific activities)
T AL/N	1.28	1.96	(10.8)	0.485 (2.7%)	163.5	0.368 (2.0%)	76.5	2.14
SVS AL/N	3.60	2.42	(13.4)	1.4 (7.8%)	162.5	1.2 (6.7%)	92.5	1.75

The glycolipid biosynthesis (CHCl$_3$-MeOH extractable material) was also diminished by the presence of GlcNAcF in the growth medium (Table 8.2). The glycolipid fraction of the control TAL/N contained 0.71% of the input ^{14}C-glucosamine counts compared with 0.31% for the SVS-AL/N cells. This is in agreement with the altered ganglioside composition of the latter reported by Mora *et al.* (1969) who showed the higher gangliosides to be missing, and GM$_3$ to be the predominant form. This has been attributed (Brady and Mora, 1970; Dijong, Mora and Brady, 1971) to depletion of hematoside UDP-N-acetylgalactosamine transferase in the SVS-strain, resulting from the viral transformation. In the presence of 0.5 mM GlcNAcF, the labelling of the glycolipids in both cell lines was markedly depressed to 0.05% of the input counts in the case of T AL/N and 0.11% for SVS AL/N cells.

Glycoprotein Biosynthesis

The foregoing experiments, concerned with the labelling of whole cells were extended to data for glycoprotein isolated from them by papain treatment and Sephadex fractionation. Description of cells and dialysis resulted in loss of about 30% of the ^{14}C-labelling,

Table 8.2

CELLS (T AL/N AND SVS AL/N) GROWN IN PRESENCE OF
0.5 mM MODIFIER FOR 16 h

Cells	^{3}H-Threonine total μCi incorporated	% of input	^{14}C-Glucosamine total μCi incorporated	% of input	^{3}H-Thymidine total μCi incorporated	% of input
GlcNAcF						
SVS AL/N						
inhibitor-free control[a]	0.14	2.9	0.36	7.2	—	—
inhibitor-free control[a]	—	—	0.42	8.3	0.34	6.8
0.5 mM-GlcNAcF[b]	1.27	2.8	3.42	7.5	—	—
T-AL/N						
inhibitor-free control[a]	0.29	5.8	0.82	16.2	—	—
inhibitor-free control[a]	—	—	1.03	20.6	0.47	9.4
0.5 mM-GlcNacF[b]	2.89	6.4	6.8	15.1	—	—
GlcNAcI						
SVS AL/N						
inhibitor-free control[a]	0.396	7.9	0.97	19.4	—	—
inhibitor-free control[a]	—	—	0.61	12.2	0.79	15.8
0.5 mM GlcNAcI[b]	4.05	9.0	9.36	20.8	—	—
T AL/N						
inhibitor-free control[a]	0.30	6.0	0.80	16.0	—	—
inhibitor-free control[a]	—	—	0.66	13.1	0.78	15.5
0.5 mM GlcNAcI[b]	3.24	7.3	12.60	27.8	—	—

[a] 5 μCi/ml of each labelled substrate 5 ml total vol., initially.

[b] 45 μCi of each labelled substrate in 45 ml initially, i.e. 9 plates each of 5 μCi.

presumably as low-molecular weight intermediates. With both
T AL/N and SVS AL/N cell lines in the presence and absence of
GlcNAcF, the proportion of ^{14}C in the diffusible fraction (low
molecular weight fraction) was in the region of 30% (Table 8.3).
In control experiments the non-diffusible glycoprotein fraction,
after 16 h growth, had a specific activity for the SVS AL/N products
of 0.20 Ci ^{14}C/mg protein, being again higher than in the corresponding T AL/N material (average 0.135 Ci ^{14}C/mg protein).
Similar results were obtained when the glycoprotein fraction was
separated by gel chromatography (Kent and Winterbourne;
also Fishman, Brady and Mora, unpublished observations). Thus

Table 8.3
^{14}C/^{3}H RATIOS FOR CELLS GROWN IN LOW CONCENTRATIONS
OF GlcNAcF LABELLED WITH (1-^{14}C) GLUCOSAMINE TOGETHER
WITH (C-^{3}H$_3$) THYMIDINE OR (^{3}H) URIDINE

	GlcNAcF (Final conc.) mM	^{14}C-Glucosamine/ ^{3}H-Thymidine	^{14}C-Glucosamine/ ^{3}H-Uridine
		(Ratios of specific activities)	
SVS AL/N	0.5	1.98	LOST
	5×10^{-2}	2.81	0.92
	5×10^{-3}	3.11	0.71
	5×10^{-4}	3.02	0.69
	5×10^{-5}	2.22	1.0
T AL/N	0.5	1.63	0.71
	5×10^{-2}	1.78	0.67
	5×10^{3}	2.02	0.52
	5×10^{-4}	2.06	0.59
	5×10^{-5}	1.86	0.58

it appears likely that a real difference exists in the rate, and possibly the type, of glycoproteins synthesized in untreated T AL/N and SVS AL/N cells, the values for the latter being elevated as a result of the virus-transformation. The findings are in keeping with observations in other cell lines that transformation by tumorigenic viruses results in extensive glycoprotein changes (Warren, Critchley and Macpherson, 1972; Buck et al., 1970, 1971; Meezan et al., 1969; Wu et al., 1970). While the general trend of our data indicates that the virally transformed line (SVS AL/N) synthesizes more glycoprotein than T AL/N, it is not possible to know with certainty at this stage whether this is stimulated synthesis of existing material or whether a glycoprotein of altered structure is formed. In a number of control experiments, the effect of N-acetylglucosamine upon the ^{14}C-glucosamine labelling of these cells was included for comparison. While under comparable conditions this sugar too modifies labelling pattern of the cells, it appears to be less effective than the halagon analogues.

Longer Term Effects of the Modifier

It is of interest that cells grown in the presence of GlcNAcF for longer periods (6 days) resulted in several-fold *increase* in specific activity of ^{14}C labelling of the glycoprotein fraction (Table 8.4). The specific activity of the SVS AL/N fraction was thus 2.1 Ci/mg protein compared to 0.26 for modifier-free control. Values for T AL/N were: 1.7 Ci/mg protein for glycoprotein fraction from cells grown in GlcNAcF, and 0.30 from modifier-free control. The

Table 8.4
EFFECT OF MODIFIERS ON THE LABELLING OF THE GLYCOLIPID AND GLYCOPROTEIN FRACTIONS

Cells and Conditions	Length of cultivation	Acetone Powder Protein mg (1)	Glycolipid fractions μCi incorporated (total)	Papain digest prior to dialysis μCi (total)	Papain digest after dialysis μCi (total)	High molecular weight fraction		
						Protein mg (total) (2)	Sp. Act. using (1)	Sp. Act. using (2)
SVS AL/N								
No modifier	1 day	66.5	0.14(0.073)[a]	15.86	10.857	—	0.163(0.24)[a]	—
0.5 mM GlcNAc	1 day	8.8	—	0.914	0.641	2.03	0.073	0.316
0.5 mM GlcNAcF	1 day	6.6	—	0.580	0.392	2.63	0.059	0.149
0.5 mM GlcNAcI	1 day	4.2	—	0.961	1.000	2.71	0.238	0.369
No modifier	6 days	—	—	0.778	0.503	1.96	—	0.257
0.5 mM GlcNAcF	6 days	—	—	3.203	2.118	1.03	—	2.056
T AL/N								
No modifier	1 day	129.5	0.32(0.129)[a]	16.87	13.03	—	0.14(0.13)[a]	—
0.5 mM GlcNAc	1 day	1.7	—	0.694	0.408	1.90	0.240	0.215
0.5 mM GlcNAcF	1 day	3.9	—	0.320	0.203	2.52	0.052	0.081
0.5 mM GlcNAcI	1 day	1.6	—	0.699	0.465	1.22	0.291	0.381
No modifier	6 days	—	—	0.760	0.467	1.55	—	0.301
0.5 mM GlcNAcF	6 days	—	—	4.819	3.104	1.82	—	1.706

[a] Results of duplicate experiments in parenthesis.

iodoanalogue (GlcNAcI) proved too inhibitory for comparable investigations.

The effect of growth in GlcNAcF appeared to bring about changes in the biosynthesis pathway without manifest alteration of growth properties of cells. While GlcNAcI displays some of the same effects it is noticeably more toxic to cells particularly in the longer term experiments. The toxic effect of GlcNAcI is to be expected on account of its formal similarity to iodoacetamide and its function has an alkylating agent which powerfully inhibits active transport and growth in *E. coli* K12 and ML 308 (Kent, Ackers and White, 1970; White and Kent, 1970; White, 1970). In *E. coli* and *B. subtilis* GlcNAcF brings about restriction of growth though it is unlikely to act by alkylation (Kent, unpublished).

N-Fluoroacetylglucosamine appears to exert little significant effect on the ^{3}H-thymidine labelling of either cell line, over a range of concentration, thus the modifier does not appear to inhibit DNA synthesis. Similarly no significant effect was shown by the modifier upon ^{3}H-uridine incorporation by either cell line suggesting that RNA biosynthesis too was not altered by the modifier (Table 8.2). The fact that these two energy-dependent pathways appear to remain unimpaired, suggests that the modifier is not acting adversely on ATP-generating processes in the cell. Preliminary metabolic experiments with mammalian epithelial cells are in keeping with this view. In other tissues, e.g. sheep colonic epithelium in vitro and calf tracheal explants in organ culture, GlcNAcF was not found to produce evident toxicity as shown by Q_{O_2} measurements, though as in the case of T AL/N and SVS AL/N cells, the compound was a significant modifier of the extent of glucosamine utilization by the tissues (Kent *et al.*, 1971).

As yet no detailed measurements of the effect, if any, of GlcNAcF upon active transport mechanisms in T AL/N or SVS AL/N cells have been made, but the similarity of labelling of soluble low-molecular weight substances in control and treated cells leads to a tentative suggestion that transport sites are not the prime points of actions of the modifiers. It has been noted that in red cells, hamster intestinal ring preparations and in rat diaphragm, GlcNAcI does not inhibit active transport of [^{14}C] methyl-D-glucoside (Kent, Ackers and White, 1970). Barnett *et al.* (1971) reported that aminosugars (glucosamine, galactosamine and mannosamine) were only weakly accumulated by hamster intestinal rings and were poor non-competitive inhibitors of the active transport of galactose by the tissue. N-acetylglucosamine was also inactive in the latter respect. The possibility that GlcNAcF undergoes metabolic transformation cannot be excluded, especially since other fluorosugars, e.g. 6-deoxy-6-fluoro-D-galactose are known to be ready substrates for enzymes (Kent and Wright, 1972). Further effects may arise from the possible incorporation of GlcNAcF in the structural elements of the cells. Evidence of such has been reported (Virijonandha and Baxter, 1970) in the cell wall structures of *Staph. aureus*. Detailed studies using ^{19}F-NMR techniques have been made of the binding of N-fluoroacetylglucosamine to specific sites in lysozyme (Dwek, Kent and Xavier, 1971; Butchard *et al.*, 1972a, b). The effect of GlcNAcF treatment on cell surface properties and on the resulting changes in biological properties of the cells, including the expression of cell surface antigens such as histocompatibility antigens, will be presented separately.

Conclusions

Some caution has to be exercised in view of the reports of variations in the biosynthetic capabilities of mammalian cells lines in the course of extended and repeated culturing. Mora (this volume, p. 64) reports apparently spontaneous changes in ganglioside composition in early and late passage T AL/N, Cowan *et al.* (1973) has also reported age-induced sequence changes in the peptides of the immunoglobulins from cultured myeloma cells. The implications of metabolic ageing and of factors which bear upon it is thus of considerable importance.

The modification of glycolipid (and glycoprotein) structure which growth in GlcNAcF implies, may be regarded as a form of experimental gangliososis. Though so far no changes have been noted in the confluence properties of such cells, the immunological and tumorigenic consequences of deliberately modified cells are dependent on the rapidity by which they are subject to anhilation by e.g. lymphocytic reaction, on the one hand, and on the other on the permanence of the modification. Recently Hughes *et al.* (1972) have shown that strain A mouse ascites tumour cells treated with neuramidase (and which exhibit decreased tumour forming ability in allogeneic mice) rapidly regenerate sialoglycoproteins at the cell surface during 6 h of subsequent culture. Further effects of neuramidase, on the cell surface, other than the removal of sialic acid residues would thus appear to expose new antigenic sites.

Nevertheless, the present investigation indicates a new and possibly fruitful approach to the question of membrane-bound membrane and the possibilities of selectively altering oligosaccharide structure during biosynthesis may repay further study.

REFERENCES

Apfel and Peters (1970) *J. Theoret. Biol.*, *26*, 47.

Barnett, J. E. G., Holman, G. D., Ralph, A. and Munday, K. A. (1971) *Biochim. Biophys. Acta*, *249*, 493.

Bekesi, J. G. and Winzler, R. J. (1969) *J. Biol. Chem.* *244*, 5663.

Bosmann, H. F. (1971) *Biochem Biophys. Acta*, *240*, 74.

Brady, R. O. and Mora, P. T. (1970) *Biochem Biophys. Acta,* *218*, 308.

Buck, C. A., Glick, M. C. and Warren, L. (1970) *Biochemistry* *9*, 4567.

Buck, C. A., Glick, M. C. and Warren, L. (1971) *Biochemistry* *10*, 2176.

Butchard, C. G., Dwek, R. A., Kent, P. W., Williams, R. J. P. and Xavier, A. (1972a) *Europ. J. Biochem.* *27*, 548.

Butchard, C. G., Dwek, R. A., Ferguson, S. J., Kent, P. W., Williams, R. J. P. and Xavier, A. V. (1972b) *FEBS Letters*, *25*, 91.

CIBA Symposium (1972) Carbon Fluorine Compounds; their chemistry biochemistry and biological activities. Assoc. Scientific Publishers, London and Amsterdam.

Cook, G. M. W. (1968) *Biol. Revs.* *43*, 363.

Cowan, N. J., Secher, D. S., Cotton, R. G. H. and Milstein, C. (1973) *FEBS Letters*, in press.

Currie, G. A. and Bagshaw, K. D. (1968) *Brit J. Cancer*, *22*, 588, 843; *23*, 141.

Dijong, I., Mora, P. T. and Brady, R. O. (1971) *Biochemistry*, *10*, 4039.

Dwek, R. A., Kent, P. W. and Xavier, A. V. (1971) *Europ. J. Biochem.* *23*, 343.

Hallinan, T., Murty, C. N. and Grant, T. H. (1968) *Arch. Biochem. Biophys.* *125*, 715.

Hughes, R. C., Sanford, B. and Jeanloz, R. W. (1972) *Proc. Nat. Acad. Sci. U.S.A.,* *69*, 942.

Hughes, R. C. (1973) *Prog. Biophys. and Mol. Biol.*, *26*, 191.

Kent, P. W., Ackers, J. P. and White, R. J. (1970) *Biochem J.* *118*, 73.

Kent, P. W., Daniel, P. F. and Gallagher, J. T. (1971) *Abs. Commun.*

Kent, P. W. (1972) *Biochem J.* *128*, 111p.

Kent, P. W. and Mora, P. T. (1973) in Biology of the Fibroblast ed E. Kulonen Academic Press, New York and London.

Kent, P. W. and Wright, J. (1972) *Carbohydrate Res.* *22*, 193.

Losick, R. and Robbins, P. W. (1967) *J. Mol. Biol.*, *30*, 445.

Marshall, R. D. and Neuberger, A. (1970) *Advanc. Carbohydrate Chem. Biochem.* *25*, 407.

Meezan, E., Wu, H. C., Black, P. H. and Robbins, P. W. (1969) *Biochemistry* *8*, 2518.

Melchers, F. (1973) *Biochemistry* in press.

Molnar, Z. and Bekesi, J. G. (1972) *Cancer Res.* *32*, 389, 756.

Mora, P. T., Brady, R. O., Bradley, R. M. and McFarland, V. W. (1969) *Proc. Natl. Acad. Sci. U.S.A.*, *63*, 1290.

Morrell, A. G., Gregoriadis, G., Scheinberg, H., Hickman, J. and Ashwell, G. (1971) *J. Biol. Chem.*, *246*, 1461.

Quastel, J. H. and Cantero, A. (1952) *Nature Lond.* 252.

Richmond, J. E., Glasser, R. M. and Todd, P. (1968) *Exptl. Cell Res.* *52*, 43.

Roseman, S. (1970) *Chem. Phys. Lipids* 5, 270.

Schenkein, I. and Uhr, J. W. (1970) *J. cell. biol.*, *46*, 42.

Virijonandha, J. and Baxter, R. M. (1970) *Biochem Biophys. Acta.* *201*, 495.

Warren, L., Critchley, D. and Macpherson, J. (1972) *Nature London* *235*, 275.

Watkins, W. M. (1970) in "Blood and Tissue Antigens" ed. D. Aminoff, Academic Press, New York and London p. 441.

White, R. J. and Kent, P. W. (1970) *Biochem. J.* *118*, 81.

White, R. J. (1970) *Biochem. J.* *118*, 89.

Winterburn, P. J. and Phelps, C. F. (1971) *Biochem. J. 121*, 701, 711, 721.
Winterburn, P. J. and Phelps, C. F. (1972) *Nature London*, 236, 147.
Winzler, R. J. (1970) *Intern. Rev. Cytol. 29*, 774.
Wu, H. C., Meezan, E., Black, P. H. and Robbins, P. W. (1969) *Biochemistry 8*, 2509.

9 Comparison of the Surface Structures of Virus-Transformed and Control Cells in Tissue Culture

L. Warren, J. P. Fuhrer and C. A. Buck

Department of Therapeutic Research, School of Medicine, University of Pennsylvania, Philadelphia, Pennsylvania and Division of Biology, Kansas State University, Manhattan, Kansas

There is at the present time considerable interest in biological membranes. Membranes are an obvious concern of workers interested in transport, virology, immunology, developmental biology, nerve function, the biochemistry of mitochondria, and DNA, RNA and protein synthesis. These investigators are deeply interested in the nature of the membrane and its role in the processes they are studying. Certainly many who have been working in the cancer field feel that the structure and behaviour of membranes and particularly the surface membrane may be fruitful objects of study to achieve an understanding of this abnormality of growth and cell division.

From a chemical point of view, components of membranes, glycolipids and glycoproteins have been investigated. Although differences in glycolipid composition between control and virus transformed (malignant) cells in tissue culture have been described (Hakomori and Murakami, 1968; Brady et al., 1969; Mora et al., 1969; Robbins and Macpherson, 1971) there are also data on cell lines which clearly show no significant differences between control and transformed cells (Warren et al., 1972a).

The glycoproteins of the surfaces of control and transformed cells have been studied by Meezan et al. (1969), Wu et al. (1969), Sakiyama and Burge (1972), Onodera and Sheinin (1970) and by our group, Buck et al. (1970, 1971a, b), Warren et al. (1972a, b, c).

We have been comparing the carbohydrate components of glycoproteins residing on the surfaces of control and virus-transformed cells in tissue culture. We have exploited the double-label technique in which control cells are grown in the presence of [^{14}C]-L-fucose and transformed cells are cultured in the presence of [^{3}H]-L-fucose. L-fucose is incorporated into glycoproteins and virtually all radioactivity can be recovered as L-fucose. After 3 days of culture in the presence of isotope during which the cells are dividing, they are treated with trypsin to free them from the glass surface upon which they had grown. This treatment removes glycopeptides which contain 20% of the bound fucose and about a quarter of the bound sialic acid of the cells (Buck *et al.*, 1970). This soluble fraction is called the trypsinate and probably consists of the more superficial components of the surface of the cell. Surface membranes are isolated from the free-floating cells by the zinc ion procedure (Warren and Glick, 1969).

The trypsinates from control and transformed cells are mixed and digested with pronase until almost all of the peptide structure is destroyed leaving intact polysaccharides which only bear the amino acid to which the carbohydrate was formerly anchored in the glycoprotein. Similarly the two sets of surface membrane, ^{14}C-labelled from control cells and ^{3}H-labelled from transformed cells are mixed dissolved in a solution of sodium dodecyl sulphate and digested with pronase. The digests are placed on columns of Sephadex G-50 (approximately 0.8×100 cm) and developed. Each tube is counted for ^{14}C and ^{3}H and the data are processed and plotted by computer. It can be seen in Fig. 9.1a, c, e that there is large, relatively early-eluting material which comes from the surface of the virus-transformed cell. A shoulder or very small peak is usually observed in this region in plots of radioactivity from control cells. Essentially the same pattern is seen using trypsinates or surface membranes so it would seem that the early-eluting material (Peak A) is present throughout the depth of the cell surface structure. The same results have been obtained when the labels are reversed i.e. control cells are grown in the presence of [^{3}H]-L-fucose and transformed cells with [^{14}C]-L-fucose.

This early-eluting material has been found in mouse, chick and hamster cells transformed by both DNA and RNA-containing oncogenic viruses (Buck *et al.*, 1970, 1971b). It is growth-dependent for the quantity of this material is reduced in both control and transformed cells in plateau phase of growth (Buck *et al.*, 1971a). Early-eluting material is found on the surface of chick embryo fibroblasts transformed by T5 virus and grown at 36° (permissive temperature) but not at 40° (Warren *et al.*, 1972a). This virus, T5 is a temperature-sensitive mutant of Rous sarcoma virus (Martin, 1970). Cells transformed by this mutant and grown at 36° appear to behave as malignant cells, while those grown at 40° are indistinguishable from controls. It is of interest that chick embryo cells transformed by T5 and grown at 36° show no significant differences

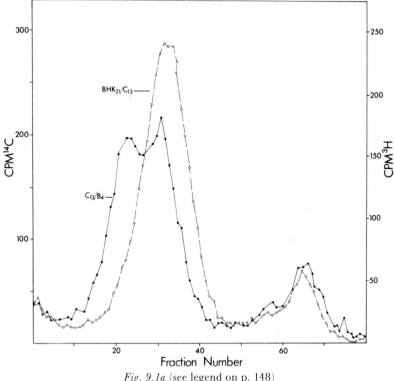

Fig. 9.1a (see legend on p. 148)

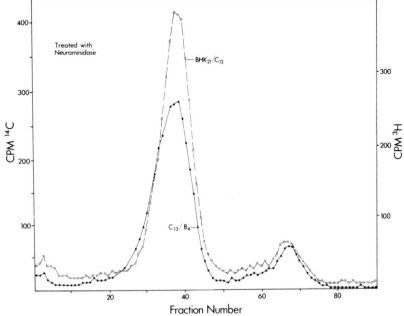

Fig. 9.1b (see legend on p. 148)

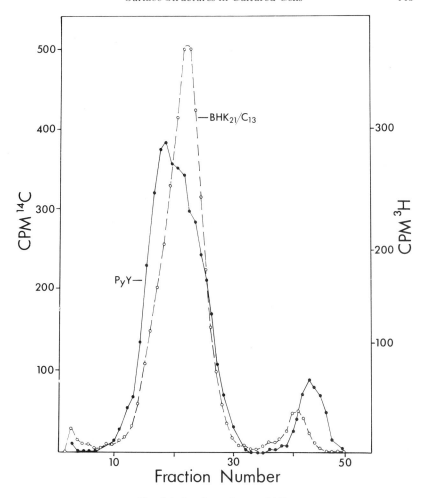

Fig. 9.1c (see legend on p. 148)

from the control, uninfected cells in their glycolipid content nor are differences observed when transformed cells are shifted from 36° (permissive temperature) to 40° (temperature at which transformation is not expressed) (Warren *et al.*, 1972a).

Since fucose residues are terminal in the small polysaccharides derived from glycoproteins, we believe that there is little degradation of these moieties during digestion with pronase. Further, it has been shown that there are very few amino acids associated with the fraction (Buck *et al.*, 1970). We know little else except that from molecular weight estimations by co-chromatography with known polysaccharide standards the glycoprotein components under study consist of

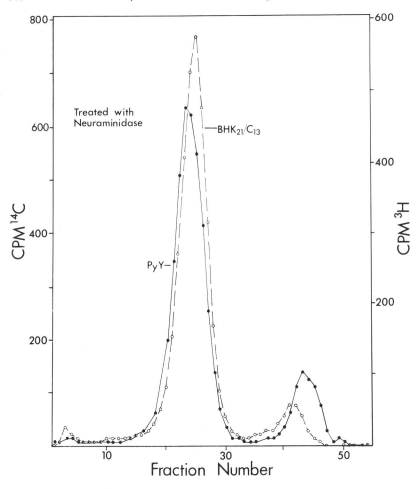

Fig. 9.1d (see legend on p. 148)

approximately 20–25 sugar residues. Some of these residues are derived from radioactive glucosamine (Buck *et al.*, 1970) but in the studies discussed here the fraction is followed by radioactivity that we know to be in the L-fucose component. The carbohydrates do not appear to combine with Concanavalin A, or with fucose-specific lectins of *Ulex europaeus* or *Lotus tetragonolobus*.

We have recently found that if pronase digests of mixed surface materials are treated with purified neuraminidase from either *V. cholerae* or *Cl. perfringens*, the large, early-eluting material (peak A) of the transformed cell disappears into a later eluting peak and the polysaccharides of control and virus-transformed cells become

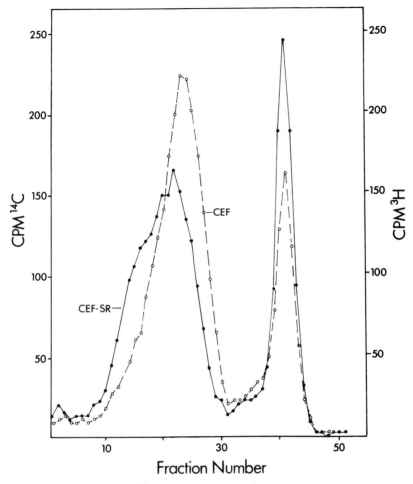

Fig. 9.1e (see legend on p. 148)

virtually the same in their elution patterns (Fig. 9.1b, d, f) (Warren
et al., 1972b, c). Similarly, if *intact* transformed cells, labelled with
radioactive fucose, are treated with neuraminidase and then with
trypsin and this trypsinate, after digestion with pronase, is cochromatographed with a pronase digest from transformed labelled
cells that had not been exposed to neuraminidase the elution pattern
seen in Fig. 9.2 is obtained. The early-eluting material (A) characteristic of transformed cells disappears into the area of the second,
large peak (B) as it did when pronase digests were treated with
neuraminidase *in vitro*. Closer examination of the elution patterns
has shown that in fact there is also a small but definite shift of peak
B to the right after treatment with neuraminidase suggesting the

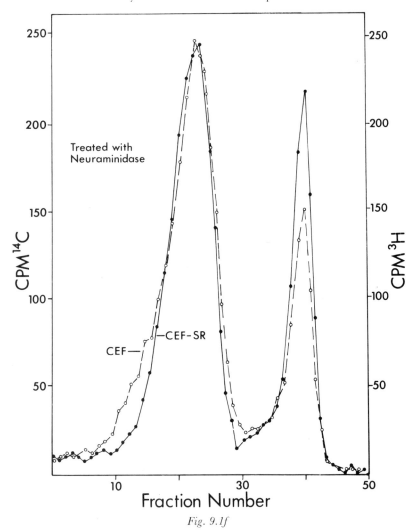

Fig. 9.1f

Fig. 9.1 Double-label elution patterns of pronase digests of trypsinates from columns of Sephadex G50 (fine). The procedure has been described in detail previously (Warren *et al.*, 1972b) and in the text. *Fig. a and b*—material derived from BHK_{21}/C_{13} labelled in log phase of growth for 3 days with $[^{14}C]$-L-fucose and from C_{13}/B_4 labelled with $[^3H]$-L-fucose. In *Fig. b* the pronase digests had been treated with 5 units of neuraminidase for 1.5 h at pH 5.2 in 3 mM $CaCl_2$ prior to application to the column. Controls were also incubated but without enzyme. *Fig. c and d*—material derived from BHK_{21}/C_{13} and its polyoma virus transformed counterpart, PyY. Note that the pattern, though slightly different from that of C_{13}/B_4 is almost identical to that previously described (Buck *et al.*, 1971b). *Fig. d*—material treated with neuraminidase before application to the column. *Fig. e and f*—material derived from chick embryo fibroblasts before and after transformation with the Schmidt-Ruppin (SR) strain of Rous Sarcoma virus. *Fig. f*—material treated with neuraminidase.

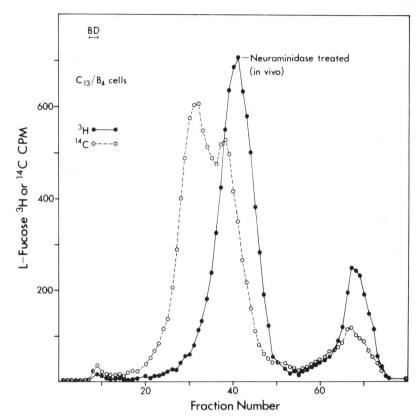

Fig. 9.2 Cells were cultured and processed as previously described (Buck *et al.*, 1970) except that cells grown in the presence of $[^{14}C]$-L-fucose were washed and incubated for one hour at 37° in the presence of Eagles minimal essential medium (MEM) without serum while cells which had been grown with $[^{3}H]$-L-fucose were incubated in MEM without serum but containing neuraminidase from *Cl perfringens*. Both cultures were then exposed to crystalline trypsin and subjected to the usual procedures.

presence of some sialic acid in the components of peak B, i.e. peak B treated with neuraminidase migrates to a peak B′ position.

An explanation for the presence of a large "peak A" in transformed cells is that these cells contain a sialyl transferase capable of transferring sialic acid (NAN) from its activated form, CMPNAN to a carbohydrate acceptor which is a component of a cell surface glycoprotein. The acceptor polysaccharide would migrate in the B′ area before the addition of NAN. After NAN is transferred, the molecule is sufficiently enlarged so that it now would elute earlier, in the A area. A considerable amount of experimental data has now been gathered to support this hypothesis (Warren *et al.*, 1972b,

1972c). At the heart of the problem is the requirement for a specific acceptor to which a hypothetical specific enzyme transfers NAN from CMPNAN.

To demonstrate the presence of a specific transferase-acceptor system, relatively large quantities of peak A material were obtained from C_{13}/B_4 (transformed) cells labelled with $[^3H]$-L-fucose. From this, sialic acid was removed enzymatically. This material, largely carbohydrate in composition and migrating in the B' area, served as acceptor. We call this acceptor "desialylated peak A".

Our assay mixture contained buffer, CMPNAN labelled with ^{14}C in the NAN moiety, acceptor (desialylated peak A) and crude, particulate extracts from control or virus-transformed cells as enzyme (Warren et al., 1972b). After incubation, blue dextran and phenol red markers were added to the mixture which was applied to a column of Sephadex G-50 (60 × 0.8 cm). Approximately 25

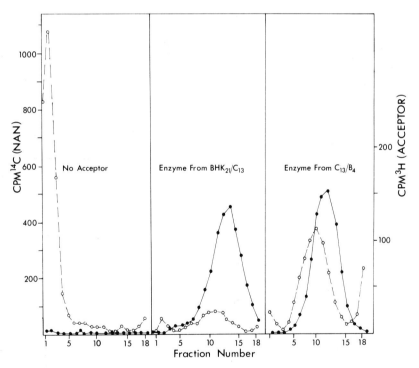

Fig. 9.3   Assay for sialyl transferase. The results of 3 assays. In the first, no added acceptor is present and there is no ^3H (acceptor) peak or ^{14}C-NAN peak in the 5th to the 15th tube region. The enzyme extract was derived from C_{13}/B_4. In the second and third assays desialylated peak A material (●—●—●) is present (^3H). O—O—O indicates transfer of ^{14}C-NAN to acceptor. 0.2 mg of particulate enzyme protein was used in each assay, and incubation was for 1 h. (See text and Warren et al. (1972b) for details.)

fractions were collected. If there has been a transfer of ^{14}C-NAN, a peak of ^{14}C (NAN) should be seen immediately preceding the ^{3}H peak of acceptor (in excess). Transfer to large acceptors, (exogenous or endogenous) can be measured by a peak of ^{14}C that elutes with the blue dextran in the void volume. These patterns can be seen in Fig. 9.3. The peak of ^{14}C-NAN in panel 3 in which the enzyme was derived from a transformed cell (C_{13}/B_4) is considerably greater than the ^{14}C peak in panel 2 where the enzyme is from the control cell (BHK_{21}/C_{13}).

Table 9.1 also shows that transfer of NAN to desialylated peak A is considerably greater in the transformed cell than in the control. However there is no detectable transfer to peak A material bearing its full complement of sialic acid. Peak B material can accept some NAN but it should be noted that NAN transferase activity to endogenous acceptor ("non specific"), peak B, desialylated fetuin or bovine mucin is the same in control and transformed cells whereas transfer to desialylated peak A is distinctly greater in the transformed cell.

Table 9.1

TRANSFER OF ^{14}C-NAN TO VARIOUS ACCEPTORS BY EXTRACTS FROM CONTROL AND VIRUS-TRANSFORMED CELLS

^{14}C-NAN Acceptor	Extract from BHK$_{21}$/C$_{13}$ cells	Extract from C$_{13}$/B$_4$ cells*
	cpm/mg protein/h	
None	0	0
Non-specific (endogenous)	1,690	2,170
Peak B	1,600	2,030
Peak A	0	0
Desialylated peak A	1,790	5,450
Desialylated fetuin	14,800	13,500
Desialylated bovine mucin	17,700	18,000

Incubation for 1 hour. (See text and Warren *et al.*, 1972b for details of assay.)
* This line of cells is derived from the hamster cell BHK$_{21}$/C$_{13}$ by transformation with a Bryan strain of Rous sarcoma virus.

To summarize our results (see Warren *et al.*, 1972b):

1. There is 2.5 to 11 times more NAN transferase activity using desialylated peak A acceptor in virus-transformed cells than in controls. However, there is activity in control cells. In several other studies of sialyl transferases in control and transformed cells activity was found to be lower in transformed cells (Grimes, 1970; Cumar *et al.*, 1970; Den *et al.*, 1971). These sialyl transferases are apparently different from the one reported here. Experiments in

which extracts of control and transformed cells are mixed does not support the notion that activity is controlled by soluble activators or inhibitors.

2. Enzyme activity is sharply reduced in both BHK_{21}/C_{13} or C_{13}/B_4 cells in plateau phase of growth as compared to rapidly dividing cells (Warren et al., 1972a). This is in agreement with the observation that peak A disappears both in control and virus-transformed cells in plateau phase of growth (Buck et al., 1971a).

3. The enzyme is found in isolated surface membranes but this is probably not its only location (Table 9.2). We now have data

Table 9.2
TRANSFER OF ^{14}C-NAN BY FRACTIONS OF CELLS IN LOG AND PLATEAU PHASE OF GROWTH

	Cell	Growth	Fraction	^{14}C incorporated into desialylated peak A material cpm/mg protein/h
a.	BHK_{21}/C_{13}	Log	Supernatant	2,140
	BHK_{21}/C_{13}	Plateau	Supernatant	546
	C_{13}/B_4	Log	Supernatant	10,850
	C_{13}/B_4	Plateau	Supernatant	1,900
b.	C_{13}/B_4	Log	crude homogenate	9,700
	C_{13}/B_4	Plateau	crude homogenate	2,860
	C_{13}/B_4	Log	surface membrane	7,400
	C_{13}/B_4	Plateau	surface membrane	2,840

1 h incubation at 37°.

(unpublished) to show that the changes upon viral transformation seen in the surface membrane of the cell described here also take place in endoplasmic reticulum (Fig. 9.4), mitochondrial membrane (inner and outer) (Fig. 9.5) and nuclear membrane. However we have not completely ruled out some contamination of these fractions by fragments of surface membranes. Experiments are in progress to rule out this contamination. If the shift does take place in internal membrane systems it might be reasonable to expect that these membranes also contain a similar NAN-transferase.

4. Chick embryo fibroblasts transformed by the temperature-sensitive mutant of Rous sarcoma virus, T5, contain considerably more of the activity (3.5 to 4 fold) when grown at 36° (permissive temperature) than at 40° at which the cells appear normal and the level is about that of control cells (Table 9.3) (Warren et al., 1972b, c).

It would appear that we are dealing with a change in the carbohydrate component of glycoprotein that is associated with

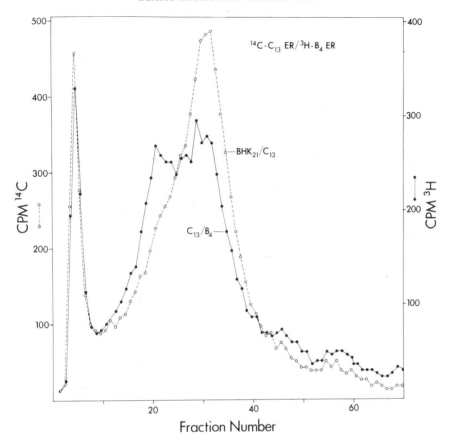

Fig. 9.4 Double-label elution patterns of pronase digests of endoplasmic reticulum from columns of Sephadex G50 (fine). BHK_{21}/C_{13} and C_{13}/B_4 cells grown in the presence of $[^{14}C]$ and $[^{3}H]$-L-fucose respectively. The procedure has been described in detail previously (Warren *et al.*, 1972b). Endoplasmic reticulum isolated from the supernatants from mitochondrial preparations by the method of Schnaitman and Greenawalt (1968). Work done in collaboration with Drs. G. Soslau and M. M. K. Nass.

membranes of virus-transformed cells. Glycoproteins of internal membrane systems as well as the surface appear to be affected. Whether the affected carbohydrate is situated on one or more species of membrane glycoprotein or whether the protein component itself is altered is unknown. We are now attempting to find out which protein(s) of the surface and other internal membranes bear the polysaccharide of peak A. It should be pointed out that the resolving power of Sephadex G-50 is relatively poor and that peaks A, B and

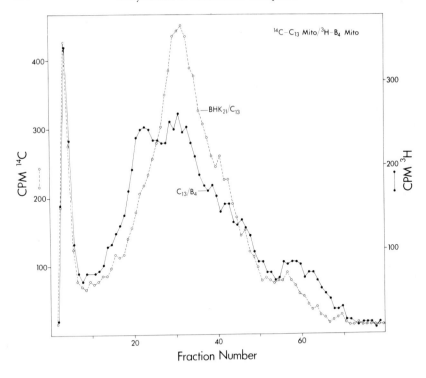

Fig. 9.5 Double-label elution patterns of pronase digests of whole mitochondria from columns of Sephadex G50 (fine). Cells used, and experimental details as previously described. Mitochondria were isolated by the procedure of Soslau and Nass (1971), and were further purified by centrifugation on a linear sucrose gradient. Work done in collaboration with Drs. G. Soslau and M. M. K. Nass.

Table 9.3
SIALIC ACID TRANSFERASE IN CHICK EMBRYO FIBROBLASTS

Cell	Temperature of culture °C	cpm of NAN-^{14}C transferred/mg protein/h	
		To endogenous acceptor	To desialyzed peak A
CEF	36	3420	4,870
CEF	41	3600	5,700
CEF-SR	36	4500	15,500
CEF-SR	41	5720	15,300
CEF-T5	36	2600	17,700
CEF-T5	41	3050	5,360

30 min incubation at 37°. 0.2 mg enzyme protein per vessel.

B′ are not homogenous. Several peaks can be resolved by DEAE-Sephadex chromatography (Fig. 9.6 and 9.7) and by high voltage paper electrophoresis. Work is progressing in identifying the components partially separated by columns of Sephadex G50 chromatography.

The difference uncovered in these studies between control and virus-transformed cells appears to be essentially quantitative in nature. "Peak A" material can be observed in both control and transformed cells (Buck *et al.*, 1970) but there is far more in the transformed cells. It would seem that an enzyme activity (sialyl transferase) associated with growth, responsible for the formation of the material in the peak A area, is present in inordinately large

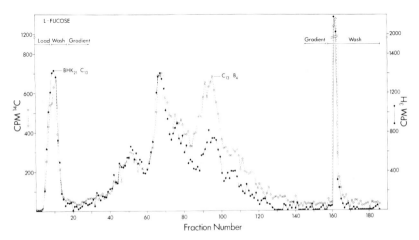

Fig. 9.6 Double-label elution patterns of pronase digests of trypsinates of $BHK_{21}/$ C_{13} and C_{13}/B_4. Column of G-25 DEAE-Sephadex (1×6 cm). Linear gradient of 0 to 0.25 M-NaCl in 0.001 M-sodium phosphate buffer, pH 6.8; total volume 120 ml. Initial wash with 0.001 M-Na phosphate buffer, pH 6.8. Final wash was with 1 M-NaCl in the same buffer. $[^{14}C]$- and $[^3H]$-L-fucose was used.

amount in transformed cells. This results in one (or possibly more) membrane glycoproteins containing extra sialic acid residues being present in excess. The transferase appears to be fairly specific and may be masked by several other sialyl transferases that do not change significantly when cells are transformed. Further, though there are extra residues of sialic acid on the glycoproteins we are concerned with, this is only a small fraction of the total sialic acid of the cell and is not a major factor in determining the *total* sialic acid level of a cell or of its membranes. Indeed there are reports that the sialic acid content of transformed cells or its membranes is lower than that of controls (Wu *et al.*, 1969). We have found that, depending

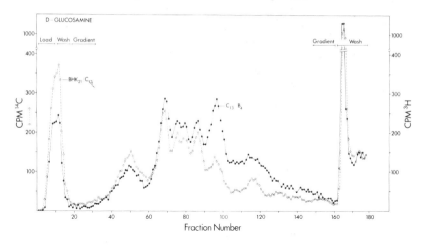

Fig. 9.7   The same procedures used except that the material used was derived from cells labelled with [^{14}C]- or [^{3}H]-D-glucosamine.

on the cell line, the sialic acid level of a virus-transformed cell may be higher, the same or lower than its normal counterpart (Hartmann *et al.*, in press).

Further work must be done to see whether this difference between control and transformed cells can be extended to other cell systems, whether the difference disappears upon reversion and whether some sort of analogous pattern is found in human, solid tumors. The consistency with which we have found this difference, the fact that it accompanies temperature sensitivity in T5 transformed cells, and its link to the growth process permits some hope that it is associated with malignancy, a disorder of growth. Our studies have focused on the cell surface where a difference has been found. This difference is consistent with modern concepts of control of cell behaviour stemming from the cell surface and altered cell behaviour (malignancy) deriving from changes in the cell surface which ultimately are the consequence of genetic alteration.

Acknowledgements

     The invaluable assistance of Mrs. Adele Gallucci, Miss Suzanne Redmond and Miss Kerstin Malmström is gratefully acknowledged. This work was supported by grants from the American Cancer Society BC-16A, PRP-28 and the U.S. Public Health Service 5 PO1 AI 0700507 and CA 12426-01.

REFERENCES

Brady, R. O., Borek, C. and Bradley, R. M. (1969). *J. Biol. Chem. 244*, 6552.

Buck, C. A., Glick, M. C. and Warren, L. (1970). *Biochemistry 9*, 4567.

Buck, C. A., Glick, M. C. and Warren, L. (1971a). *Biochemistry 10*, 2176.

Buck, C. A., Glick, M. C. and Warren, L. (1971b). *Science 172*, 169.

Cumar, F. A., Brady, R. O., Kolodny, E. H., McFarland, V. W. and Mora, P. T. (1970). *Proc. Nat. Acad. Sci. U.S.A. 67*, 757.

Den, H., Schultz, A. M., Basu, M. and Roseman, S. (1971). *J. Biol. Chem. 245*, 2721.

Grimes, W. J. (1970). *Biochemistry 9*, 5083.

Hakomori, S. and Murakami, V. T. (1968). *Proc. Nat. Acad. Sci. U.S.A. 59*, 254.

Hartmann, J. F., Buck, C. A., Defendi, V., Glick, M. C. and Warren, L. *J. Cell Physiol.* in press.

Martin, G. S. (1970). *Nature (London) 227*, 1021.

Meezan, E., Wu, H. C., Black, P. H. and Robbins, P. W. (1969). *Biochemistry 8*, 2518.

Mora, P. T., Brady, R. O., Bradley, R. M. and McFarland, V. W. (1969). *Proc. Nat Acad. Sci. U.S.A., 63*, 1290.

Onodera, K. and Sheinin, R. (1970). *J. Cell Sci. 7*, 337.

Robbins, P. W. and Macpherson, I. (1971). *Proc. Roy. Soc. B 177*, 49.

Sakiyama, H. and Burge, B. W. (1972). *Biochemistry 11*, 1366.

Soslau, G. and Nass, M. M. K. (1971). *J. Cell Biol. 51*, 514.

Warren, L., Critchley, D. and Macpherson, I. (1972a). *Nature (London) 235*, 275.

Warren, L., Fuhrer, J. P. and Buck, C. A. (1972b). *Proc. Nat. Acad. Sci. U.S.A. 69*, 1838.

Warren, L., Fuhrer, J. P. and Buck, C. A. (1972c). *Federation Proc.* in press.

Warren, L. and Glick, M. C. in *Fundamental Techniques in Virology*, Eds. K. Habel and N. P. Salzman, Academic Press, New York, p. 66, 1969.

Wu, H. C., Meezan, I., Black, P. H. and Robbins, P. W. (1969). *Biochemistry 8*, 2509.

10 *Membrane Teichoic Acids and Their Interaction with the Bacterial Membrane*

I. C. Hancock and J. Baddiley

School of Chemistry, The University, Newcastle-upon-Tyne, England

Teichoic acids occur at two locations in the cells of Gram-positive bacteria. The "wall teichoic acid" may have a wide variety of different structures whose only common feature is the presence of polyol phosphate residues in the polymer chain, and is covalently linked to the peptidoglycan of the cell wall. The "membrane teichoic acid" is always a linear polymer of glycerol phosphate residues, to one end of which is attached a glycolipid which is identical with the normal membrane glycolipid. This "lipoteichoic acid" is attached to the outer surface of the cell membrane by some form of interaction involving magnesium ions and also, presumably, lipophilic forces. The phosphate groups of the membrane teichoic acid bind a stoichiometric amount of Mg^{++}, whose presence affects the response of Mg-dependent enzymes in the membrane to varying concentrations of the divalent cation in aqueous solution. Thus, the dependence of the biosynthesis of wall polymers by membrane-bound enzymes upon Mg^{++} is severely "damped" by the presence of Mg-lipoteichoic acid so that the enzymes exhibit near-maximal activity over a wide range of concentrations of added magnesium ions. When the membranes are surrounded by cell walls the wall teichoic acid acts as a reservoir of bound Mg^{++} and the membrane bound enzymes become completely insensitive to externally added Mg^{++}. The evidence suggests that the membrane, or membrane-bound enzymes, interact directly and preferentially with Mg^{++} which is bound to the membrane teichoic acid.

Teichoic acids are strongly acidic, phosphate-containing polymers which occur in various amounts in all Gram-positive bacteria. There is a great diversity of structural types whose only common feature is the presence of polyol phosphate residues, either ribitol phosphate or glycerol phosphate, in the backbone of the polymer. Some representatives of the various types are shown in Fig. 10.1. Hexoses often occur in these polymers, sometimes as

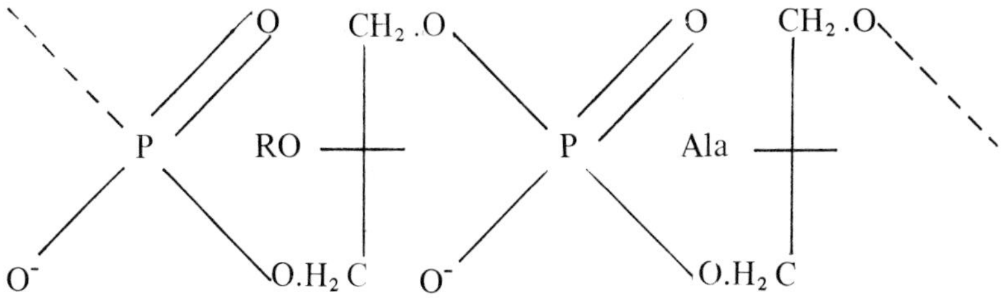

R = glycosyl

poly (glycerol phosphate)

B. licheniformis ATCC 9945

    [-3 glycerol 1-phosphate-6 glucose α 1-]$_n$

S. lactis 13

    [-3 glycerol 1-phosphate-4N-acetylglucosamine α 1-phosphate-]$_n$

Fig. 10.1 Teichoic acid structures.

glycosyl substituents on the hydroxyl groups of the polyols, or alternatively as components of the polymer backbone, where they may be linked glycosidically or through a phosphodiester group to the polyol phosphates. D-Alanine has been found in most of the polymers whose structures have been determined; it is esterified either to the polyol or to the glycosyl substituents. The structural variants are not species-specific. For example, representatives of several genera contain poly(glycerol phosphate) substituted with glucose; on the other hand three strains of the species Bacillus subtilis contain respectively poly(glycerol phosphate), poly(ribitol phosphate) and poly(glycerol phosphate glucose) (Baddiley, 1972).

The teichoic acids are found at two distinct locations in the bacterial cell: in the cell wall, where they are covalently linked to the peptidoglycan network, and at an intracellular site. The two groups of polymers often differ in structure. Thus, while the wall teichoic acid may have any of the structures shown in Fig. 10.1, the intracellular teichoic acid is always a derivative of poly(glycerol phosphate). The intracellular teichoic acids attracted attention when Wicken, Eliott and Baddiley (1963) showed that the Group D antigen of Streptococci was the intracellular polymer. Attempts to identify the precise location of the antigen in the cell showed that it occupied a region between the cytoplasmic membrane and the cell wall (Hay et al., 1963). When the cell walls were digested with lyso-zyme in a hypertonic medium to yield intact protoplasts a large proportion of the antigen was released into solution; the rest was found in the particulate membrane fraction after lysis of the proto-plasts (Shockman and Slade, 1964). The proportions of the polymer bound to the protoplast membrane and released into solution appeared to depend on the composition of the suspension medium and in particular a high concentration of magnesium ions (10 mM) caused retention of most of the teichoic acid on the membrane. The same phenomenon has been observed in a number of strains of Gram-positive bacteria and the intracellular polymer has therefore come to be called "membrane teichoic acid".

Membranes prepared in the presence of Mg^{++} and therefore containing bound membrane teichoic acid could be extensively washed with water or saline without release of all the teichoic acid. It was therefore suggested that the poly(glycerol phosphate) chain might be covalently linked to some component of the membrane (Archibald et al., 1968). Early chemical studies did not reveal such a linkage, but they were carried out on polymer which had been extracted with trichloroacetic acid which might have degraded the native material. The conclusion that the structure of membrane teichoic acid was more complex than that indicated by studies of TCA-extracted material was reinforced by observations of the effect of aqueous phenol on membranes. Burger and Glaser (1964) observed that 45% aqueous phenol efficiently extracted membrane teichoic acid into aqueous solution and they noted that polymer prepared in this way had a much higher apparent molecular weight than did acid-extracted material. Wicken and Knox (1970) used the phenol technique to isolate the membrane teichoic acid of Lacto-bacillus fermenti and they too found that the product had an extremely high apparent molecular weight. They concluded from the nature of the products of alkaline hydrolysis of the polymer that it was combined with a glycolipid and they gave it the name "lipoteichoic acid". This complex was antigenic when injected into rabbits with adjuvant, whereas the lipid-free teichoic acid, extracted with TCA, was not. Using similar methods Toon et al. (1972) examined in detail the membrane teichoic acid of a Group D streptococcus and found that the phenol-extracted polymer possessed a chain of about thirty units of glycerol phosphate, partially substituted with glucosyl (kojibiosyl) residues and covalently linked through its terminal phosphate group to a diglucosyl diglyceride identical with the glycolipid found in the membrane. The presence of the lipid gave the molecule amphiphilic properties which caused micelle formation in aqueous solution, thus giving rise to the high apparent molecular weight observed on gel-filtration or ultracentrifugation. Removal of the fatty acids with hydroxylamine led to the breakdown of the micellar structure and a great reduction in the measured molecular weight. Coley *et al.* (1972) have examined a range of bacteria representative of several genera and have demonstrated that in each case the membrane teichoic acid is a lipoteichoic acid. Their detailed work has also revealed the interesting fact that each lipoteichoic acid molecule contains two residues of N-acetylglucosamine as substituents at the 2-positions of two of the glycerol residues of the polymer chain. Although their precise location is unknown it seems likely that they occupy specific sites, perhaps marking one end of the anionic part of the molecule. The general features of the structure of lipoteichoic acid are shown in Fig. 10.2.

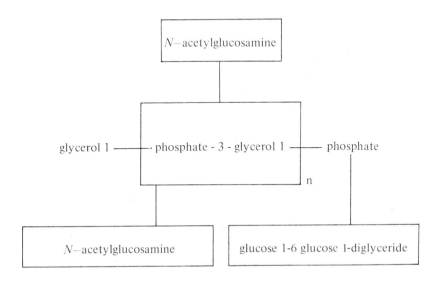

Fig. 10.2 Lipoteichoic acid

LIPOTEICHOIC ACID AND THE CYTOPLASMIC MEMBRANE

At least two factors are involved in the interaction between the outer surface of the cytoplasmic membrane and the lipoteichoic acid: the binding of divalent cations to the phosphate groups of the polymer and the hydrophobic properties of the glycolipid group. As has already been mentioned, most of the lipoteichoic acid remains attached to the membrane when the cell wall is removed only if a high concentration of Mg^{++} (10 mM) is maintained during the process. Membranes which have been prepared under these conditions can then be washed until a constant amount of teichoic acid remains bound. Table 10.1 shows the composition of such a prepara-

Table 10.1

AMOUNTS OF LIPOTEICHOIC ACID AND BOUND Mg^{++} IN
PREPARATIONS OF PROTOPLAST MEMBRANES FROM
B. LICHENIFORMIS

Preparation	lipoteichoic acid (μmol phosphate/ mg dry weight)	bound Mg^{++} (μequiv/mg dry weight)
A (prepared in the absence of Mg^{++})	0.05	0.056
B (prepared in the presence of Mg^{++})	0.43	0.38
A + bound lipoteichoic acid	0.42	0.38
B after washing with EDTA	0.12	0.026
C (wall membrane)	0.47	1.42
	(Total TA 1.3)	

tion from *Bacillus licheniformis* A.T.C.C. 9945. The binding of Mg^{++} to the phosphate groups of the polymer is stoichiometric: under no conditions does the ratio of Mg^{++} (equivalents) to phosphate groups significantly exceed unity. This indicates that there can be very little 'bridging' of magnesium ions between polymer phosphate groups and ionic groups in the membrane since that would give ratios of Mg/P above 1, approaching 2 in the limiting case. Three types of interaction between polymer, magnesium ions and membrane remain possible. Firstly, Mg^{++} may be bound to the phosphate groups of an individual polymer chain, without any direct interaction with other chains or with the membrane; such bound cations could facilitate more compact folding of the chain by electrostatic shielding of the charged phosphate groups. A second possibility is that some of the teichoic acid may be attached to the membrane-lipoteichoic acid complex only by cation bridges to the phosphate groups of the membrane-bound lipoteichoic acid.

Thirdly, a small proportion of the total phosphate groups of the bound lipoteichoic acid may interact with ligand groups in the membrane through Mg^{++} bridges. Table 10.2 shows that by washing

Table 10.2
EFFECT OF WASHING WITH SODIUM EDTA ON THE AMOUNTS OF BOUND Mg^{++} AND TEICHOIC ACID IN PROTOPLAST MEMBRANE

	teichoic acid (μmol phosphate/mg dry weight)	bound Mg^{++} (μequiv/mg dry weight)
Original preparation	0.37	0.25
1 wash with EDTA	0.24	0.044
2 washes with EDTA	0.15	0.026

membrane-lipoteichoic acid preparations with EDTA a large proportion (84%) of the total bound Mg^{++} can be removed with the loss of only 35% of the teichoic acid. However, removal of more Mg^{++} then leads to a more rapid loss of teichoic acid. This effect is shown in more detail in Fig. 10.3. It appears that some of the Mg^{++} is

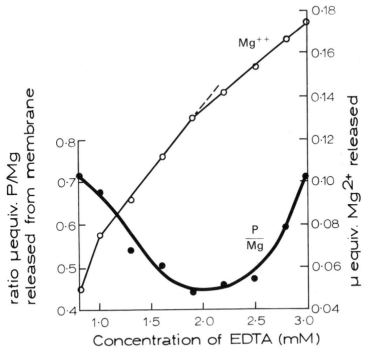

Fig. 10.3 Effect of Mg^{2+} removal on teichoic acid release from membrane preparations.

rather firmly bound and can only be released with difficulty at high EDTA concentrations, as indicated by the lower gradient of the upper part of the Mg-removal curve. Loss of this Mg^{++} is accompanied by a sudden increase in the release of teichoic acid accompanying the removal of a fixed amount of Mg^{++}. Possibly the more firmly bound Mg^{++} is involved in bridging between a small number of phosphate groups in the teichoic acid and some ligand in the membrane which is essential for attachment of the lipoteichoic acid to the membrane.

The binding of the paramagnetic ion Mn^{++} to randomly coiled, single strand polyribonucleotides has been studied by Eisinger *et al.* (1965), and Crutchfield and Irani (1965) have investigated the interaction of a number of cations with condensed phosphate polymers. The general conclusions drawn from their work are that divalent cations are site-bound at the phosphate groups by "strong interactions" which take the binding to completion in the presence of a very small excess of cations. Similar studies have yet to be carried out on teichoic acids, but the general principles seem to be the same.

The glycolipid moiety of the lipoteichoic acid must play an important role in the attachment of the polymer to the membrane. The lipid can be envisaged intercalated with the membrane lipids, with its fatty acid chains buried in the hydrophobic interior of a lipid bilayer. Under these conditions the lipid could readily adopt a conformation with the hydroxyl groups of the hexose rings arrayed on the outer surface of the membrane, attached to the poly(glycerol phosphate) chain. This is very similar to the mode of attachment of lipopolysaccharide to phospholipid monolayers which has been studied by Rothfield and his co-workers (Rothfield *et al.*, 1972). In their system, divalent cations were not essential for the recombination of polymer and phospholipid, although subsequent adsorption of an enzyme to form a functional ternary complex was Mg-dependent.

We have examined the adsorption of purified lipoteichoic acid to membrane vesicles prepared from *B. licheniformis*. Fig. 10.4 shows that some adsorption occurs in the absence of added Mg^{++} but this is approximately doubled at 10 mM $MgCl_2$. Table 10.1 compares the teichoic acid and Mg^{++} contents of the reconstituted system with those of membranes prepared by lysis of cells in the presence of 10 mM $MgCl_2$, showing that the two systems are almost identical. This is perhaps an indication that there is a limited number of binding sites for polymer on the membrane.

There is considerable evidence that membrane teichoic acids are essential for normal cellular activity in Gram-positive bacteria.

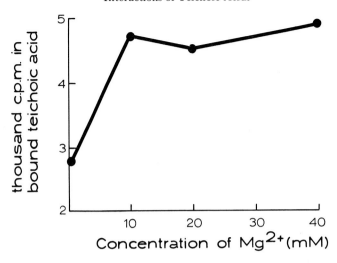

Fig. 10.4   Requirement for Mg^{2+} of the binding of lipoteichoic acid to membrane.

They have been detected in all species so far examined and are invariably polymers of the poly(glycerol phosphate) type; recent work suggests that they are all lipoteichoic acids. When *B. subtilis* is grown in a chemostat under conditions of phosphate limitation the teichoic acid in its wall is replaced by an acidic polysaccharide which does not contain phosphate (Tempest *et al.*, 1968) but the normal membrane teichoic acid continues to be synthesized (Ellwood and Tempest, 1968). The evidence suggests that the teichoic acids play a part in the accumulation of divalent cations by the cell. Because of the strong interaction between phosphate groups and divalent cations, cell walls containing teichoic acids can 'scavenge' the cations even when these are present at a very low concentration in the growth medium. It has been found that *B. subtilis* grown in a chemostat at a limiting concentration of Mg^{++} synthesizes cell walls containing abnormally large quantities of teichoic acid (Tempest *et al.*, 1967). This response would enable the cell walls to accumulate the available divalent cations more readily. The presence of membrane teichoic acid between the cell wall and the cytoplasmic membrane could provide an integrated ion-exchange system from the exterior of the cell through the wall to the surface of the membrane. There, high concentrations of Mg^{++} are required for membrane stability (Weibull, 1967; Reavely and Rogers, 1969) and a variety of enzymic reactions, particularly those responsible for the biosynthesis of the membrane and the cell wall. Thus in bacilli the synthesis of protein, phospholipid (Patterson and Lennartz, 1971), teichoic acid (Burger and Glaser, 1964) and peptidoglycan (Anderson *et al.*, 1972) take place on the membrane and are all dependent on Mg^{++}. The optimal concentrations of Mg^{++} for the isolated biosynthetic systems vary from 8 mM (for phospholipid) to 60 mM (for peptidoglycan) with most requirements falling in the range 10–30 mM. These facts led us to investigate the effects of teichoic acids on the response to divalent cations of some of the membrane-bound enzyme systems. (Hughes *et al.*, 1973.)

GLYCEROL PHOSPHATES

We first examined the Mg-dependence of the enzymes responsible for the biosynthesis of poly(glycerol phosphate) and poly-(glycerol phosphate glucose) in isolated membrane vesicles obtained by the lysis of *B. licheniformis* with lysozyme. Some details of the membrane preparations are shown in Table 10.1. Membrane A was prepared in the absence of Mg^{++} and contained almost no lipoteichoic acid or bound Mg^{++}. Preparation B was obtained in the presence of Mg^{++} and contained eight times as much lipoteichoic acid with a stoichiometric amount of bound Mg^{++}.

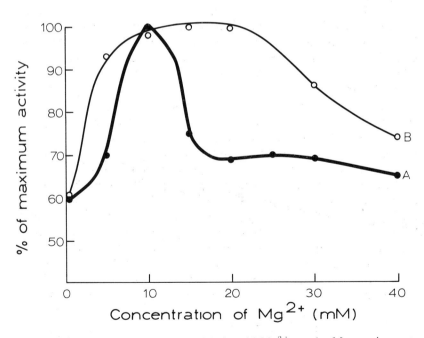

Fig. 10.5 The effect of bound lipoteichoic acid-Mg^{2+} on the Mg requirement for the biosynthesis of poly (glycerol-phosphate-glucose).

Fig. 10.5 shows that in preparation B the enzymes exhibited 90% or more of their maximum activity over a range of Mg^{++} concentrations of 20 mM whereas preparation A showed a very sharp dependence on Mg^{++}, with near-maximal activity over a range of only 5 mM. An identical effect was observed when lipoteichoic acid was adsorbed to preparation A in the presence of 10 mM Mg^{++}: the reconstituted system behaved exactly like preparation B, and contained the same amounts of bound lipoteichoic acid and Mg^{++}. Fig. 10.6 shows the effect of adsorbed lipoteichoic acid on a

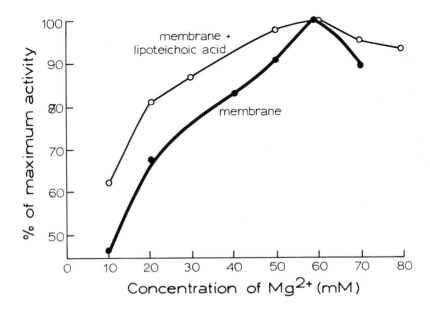

Fig. 10.6 The effect of bound lipoteichoic acid on the Mg requirement for biosynthesis of peptidoglycen by membranes of *B. licheniformis*.

membrane preparation catalysing the synthesis of peptidoglycan. Again, the lipoteichoic acid "damped" the response of the enzyme to added Mg^{++}. When the lipoteichoic acid and bound Mg^{++} were removed from preparation B or the reconstituted system using EDTA, the Mg-dependence of the enzymes reverted to those of preparation A.

It is clear that bound lipoteichoic acid modifies the requirement of these membrane bound enzymes for externally added Mg^{++}. The effect is most pronounced at high concentrations of Mg^{++}, where the presence of polymer reduces inhibition by the cations.

This can not be due to complexing of added cations with the teichoic acid since the phosphate groups are already saturated with Mg^{++} and the results therefore suggest that the magnesium ions already bound to the lipoteichoic acid on the membrane surface can activate the membrane-bound enzyme systems, and that at high concentrations of externally added Mg^{++} the enzymes interact preferentially with the bound cations. The possibility that in whole cells the wall teichoic acid acts as a "reservoir" of bound divalent cations for the lipoteichoic acid-membrane complex was investigated using preparations of broken cells in which the membrane was still vesicular in form and carried adsorbed lipoteichoic acid but, this time, was surrounded by almost complete cell wall. Such a preparation could be obtained by breaking cells of *B. licheniformis* in a French Pressure Cell. Fig. 10.7 shows the Mg-dependence curve of such a preparation

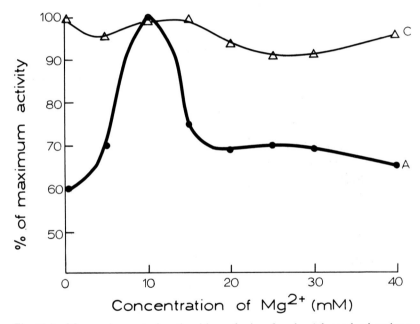

Fig. 10.7 Mg-requirement for the biosynthesis of poly (glycerol phosphate glucose) by a wall membrane preparation.

for teichoic acid biosynthesis. In this case the enzymes were almost completely insensitive to externally added Mg^{++}, but on removal of the walls by digestion with lysozyme they reverted to the state observed for isolated membranes. Both the wall and membrane teichoic acids were saturated with bound Mg^{++} (see Table 10.1) and the wall teichoic acid had about twice the binding capacity of the lipoteichoic acid.

All the results obtained to date support the conclusion that membrane bound lipoteichoic acid provides the correct environment of Mg^{++} for the cytoplasmic membrane and that the membrane teichoic acid in turn preferentially interacts with Mg^{++} in the pool bound to the wall teichoic acid, rather than with the ions in solution. There is indirect evidence that the membrane teichoic acid interacts with enzymes in the periplasmic space between the membrane and the wall, as well as with the membrane-bound enzymes. Autolytic enzymes have been extracted from the cell associated with teichoic acid (Brown et al., 1970), and under certain conditions high concentrations (5 M) of NaCl can cause release (Brown, 1972) or activation (Gilpin et al., 1972) of autolytic enzymes. Activation of autolytic enzymes by NaCl in a mutant of Staphylococcus aureus deficient in wall teichoic acid is inhibited by $MgCl_2$ (Gilpin et al., 1972). These results are consistent with the possibility of cation interactions between the autolytic enzymes and teichoic acid in the wall or periplasmic space. This may be another way in which the membrane teichoic acid plays a part in the control of the synthesis of the surface layers of bacteria.

The work which we have discussed supports the hypothesis that membrane teichoic acid is concerned in the control of divalent cation-dependent processes in the outer layers of the bacterial cell and that in many cases teichoic acid covalently linked to the wall can act as a reservoir for phosphate-bound cations with which the membrane teichoic acid can directly interact.

REFERENCES

Anderson, R. G., Hussey, H. and Baddiley, J. (1972) Biochem. J. 127, 11–25.

Archibald, A. R., Baddiley, J. and Blumsom, N. L. (1968) Adv. in Enzymol. 30, 223–253.

Baddiley, J. (1972) Essays in Biochemistry, Ed. Campbell, B. N. and Dickens, F. Vol. 8, 35–79.

Brown, W. C. (1972) Biochem. Biophys. Res. Commun. 47, 993–996.

Brown, W. C., Fraser, D. K. and Young, F. E. (1970) Biochim. Biophys. Acta 198, 308–315.

Burger, M. M. and Glaser, L. (1964) J. Biol. Chem. 239, 3168–3177.

Coley, J., Duckworth, M. and Baddiley, J. (1972) J. Gen. Microbiol. 73, 587–591.

Crutchfield, M. M. and Irani, R. R. (1965) J. Am. Chem. Soc. 87, 2815–2819.

Eisinger, J., Fawaz-Estrup, F. and Shulman, R. G. (1965) J. Chem. Phys. 42, 43–48.

Ellwood, D. C. and Tempest, D. W. (1968) Biochem. J. 108, 40P.

Gilpin, R. W., Chatterjee, A. N. and Young, F. E. (1972) J. Bacteriol. 111, 272–283.

Hay, J. B., Wicken, A. J. and Baddiley, J. (1963) Biochem. Biophys. Acta 71, 188–190.

Hughes, A. H., Hancock, I. C. and Baddiley, J. (1973) Biochem. J. 132, 83–93.

Patterson, P. H. and Lennartz, W. J. (1971) J. Biol. Chem. 246, 1061–1072.

Reavely, D. A. and Rogers, H. J. (1969) *Biochem. J. 113*, 67–79.
Rothfield, L., Romeo, D. and Hinckley, L. (1972) *Fed. Proc. 31*, 12–14.
Shockman, G. D. and Slade, H. D. (1964) *J. Gen. Microbiol. 37*, 297–305.
Tempest, D. W., Dicks, J. W. and Meers, J. L. (1967) *J. Gen. Microbiol. 49*, 139–145.
Tempest, D. W., Dicks, J. W. and Ellwood, D. C. (1968) *Biochem. J. 106*, 237–243.
Toon, P., Brown, P. E. and Baddiley, J. (1972) *Biochem. J. 127*, 399–409.
Weibull, C. (1967) *Symp. Soc. Gen. Microbiol. 6*, 111–126.
Wicken, A. J., Eliott, S. D. and Baddiley, J. (1963) *J. Gen. Microbiol. 31*, 231–237.
Wicken, A. J. and Knox, K. W. (1970) *J. Gen. Microbiol. 60*, 293–302.

11 *Phospholipid Turnover as a Determinant of Membrane Function*

C. A. Pasternak and P. Knox
Department of Biochemistry, University of Oxford, Oxford OX1 3QU, U.K.

Phospholipids are generally thought of as "bulk", non-specific constituents of membranes. In the sense that phospholipid composition can be varied considerably without apparent effect on the properties of membranes, this is true. On the other hand, it is becoming clear that many membrane-bound enzymes require a particular phospholipid for maximal activity and that specific changes in phospholipid metabolism accompany certain physiological events. Hence, it is pertinent to ask whether *minor* changes in phospholipid content are associated with membrane-mediated information transfer.

Phospholipids turn over extensively in most cell types (Dawson, 1966). Turnover occurs irrespective of whether cells are dividing or static, whether in culture or animals, whether neoplastic or normal (Pasternak and Bergeron, 1970; Pasternak and Friedrichs, 1970; Pasternak, 1972, 1973c). The rate of turnover does not increase as cells stop growing or lose viability; on the contrary, the rate is maximal when cells are at the peak of their biosynthetic activity (Bergeron *et al.*, 1970; Warmsley and Pasternak, 1970) and is greater in healthy than in deficient animals (Pasternak and Friedrichs, 1970). In other words, any change in phospholipid turnover accompanying information transfer would be superimposed on a relatively high "background" rate. But why should an alteration in rate of turnover, which is the result of synthesis and degradation of a particular molecular species, lead to any *net* changes in phospholipid composition at all? Where the same molecule is subject to concomitant degradation and resynthesis, it clearly does not. But where degradation and resynthesis are separated in space or in time, the pattern of phospholipids in the membrane may be caused to vary. Since the situations in which phospholipid turnover is increased are often associated with some kind of membrane rearrangement (Pasternak, 1973a; Pasternak and Quinn, 1973), the latter possibility is a likely one. It is strengthened by the observation that turnover may be restricted by certain spatial considerations (Pasternak and Micklem, 1973).

Two cases in which increased turnover have been reported are particularly pertinent to the transmission of information: the activation of lymphocytes (Ling, 1968) and the resumption of growth in contact-inhibited cells (Dulbecco, 1971; Stoker, 1972). In each case cells in the "resting", G_0 phase are stimulated to re-enter the cell cycle, and to proceed, through G_1, S and G_2, to cell division. The other point in common is that the information is mediated from outside via the cell surface. The trigger is generally an antigen or mitogen in the case of lymphocytes, and a raised serum concentration in the case of contact-inhibited cells. However, other substances seem able to initiate at least some of the steps leading to cell division in each case, and it may be that these early effects are not as specific as originally thought. Certainly with regard to the progression of events leading to malignancy—with which serum-mediated relief of contact inhibition has some similarity—it is the *subsequent* failure to re-commit a cell to G_0, rather than the *initial* relief from G_0, that appears to be defective (Pasternak, 1973b).

When lymphocytes are activated by phytohaemagglutinin, phospholipid synthesis is stimulated (Fisher and Mueller, 1968, 1969) and this reflects increased turnover, since degradation is stimulated at the same time (Pasternak and Friedrichs, 1970). Again, the increased phospholipid synthesis that accompanies serum-stimulated relief of contact-inhibited cells (Cunningham and Pardee, 1969, 1971; Cunningham, 1972) is largely due to turnover (Pasternak, 1972, 1973c). In order to clarify the effect of serum the following experiments were carried out.

First, it seemed desirable to study the effects of serum in suspension culture, rather than on plates, in order to improve the reproducibility of cell sampling. Hepatoma tissue culture (HTC) cells (Tomkins *et al.*, 1966), kindly donated by Dr. G. M. Tomkins and by Dr. Bridgid Hogan, were used. Cells, grown for four days on plates in the presence of radioactive precursors were washed, transferred to suspension medium (10^5/ml) (half-strength Dulbecco medium buffered with tricine) without serum and stirred with a magnetic bar overnight. Foetal calf serum was then added (the control culture received 5% albumin in saline) and samples removed for assay at intervals. Increasing concentrations of serum caused increasing stimulation of phospholipid turnover (Fig. 11.1). That serum indeed has a growth-promoting effect on these cells is shown by monitoring certain changes, some of which are shown in Fig. 11.2.

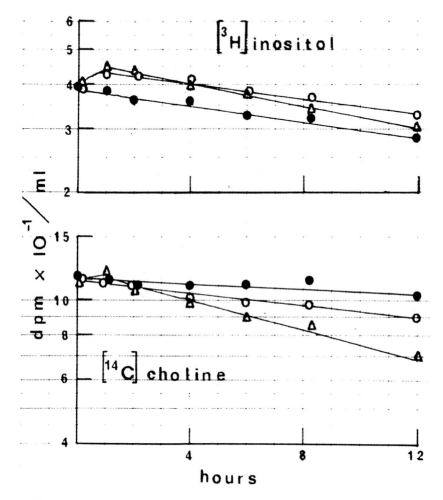

Fig. 11.1   Effect of serum on phospholipid turnover in HTC cells. HTC cells exposed to [³H]inositol (1.2 μCi/ml) and (1,2-¹⁴C)choline (0.2 μCi/ml) for four days were washed, trypsinized and maintained in serum-less modified Dulbecco medium in spinner culture for 16 h. The culture was divided into three parts and one hour later exposed to 5% albumin (control), ●—●; 5% foetal calf serum, O—O; or 15% foetal calf serum, △—△. Samples were removed at intervals and radioactivity incorporated into phospholipid measured (Pasternak and Bergeron, 1970).

SPECIFIC SERUM STIMULATED EVENTS

OF HTC CELLS IN SUSPENSION CULTURE

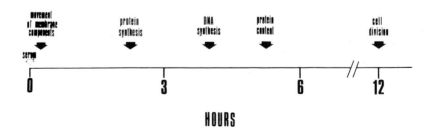

HOURS

Fig. 11.2 Sequence of events following addition of serum to HTC cells. The arrows refer to the approximate times at which events become detectable.

The timing of these events is more gradual than is apparent from Fig. 11.2, and varies somewhat from experiment to experiment; the arrows in Fig. 2 represent the mean of several experiments. Although cells at 10^5/ml are far below their saturation density, let alone a state of contact-inhibition, they *do* appear to be in G_0, and the sequence of events following serum stimulation is just that seen in other situations. Moreover, the early events now to be described have also been observed in confluent 3T3 cells, which *are* in a state of contact-inhibition.

The changes in phospholipid precursors, or breakdown products, was next studied. When pre-labelled HTC cells are exposed to serum, some isotope rapidly appears in the medium (Fig. 11.3). This is not due to gross damage of the plasma membrane, since previously-accumulated $^{51}CrO_4^{--}$, for example, does not leak out. Nor does isotope from cells pre-labelled with valine appear (Fig. 11.6), though changes in uptake of amino acids (Wiebel and Baserga, 1969) and other compounds has been reported (Cunningham and Pardee, 1969; Sefton and Rubin, 1971; Vaheri *et al.*, 1972). Differences in cell type, condition of culture, and other experimental details such as measurement of efflux as opposed to uptake, are likely to account for any discrepancies.

The appearance of isotopic labels in the medium is not matched by loss from the intracellular soluble pool. On the contrary, this too rises shortly after the addition of serum (Fig. 11.4). Analysis of this fraction by paper chromatography reveals, in the case of choline-labelled cells, three main constituents (Fig. 11.5).

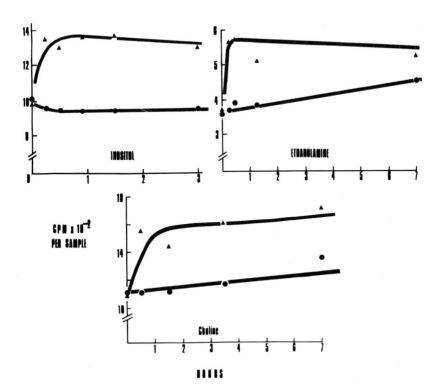

Fig. 11.3   Effect of serum on the appearance of isotope in the medium. HTC cells exposed to [³H]inositol (1.2 µCi/ml), [³H]ethanolamine (0.3 µCi/ml) or [Me-³H]-choline (0.2 µCi/ml) for four days were treated as in Fig. 11.1 and divided into two parts prior to the addition of 5% albumin (control) ●—●, or 10% foetal calf serum ▲—▲. Radioactivity in the medium was assayed following removal of the cells by centrifugation.

These have been tentatively identified as phosphorylcholine, choline and phosphatidylcholine. The fact that most of the non-lipid material is in the form of phosphorylcholine rather than choline agrees with an earlier report (Plagemann, 1971). The presence of phosphatidylcholine in this fraction is surprising, and appears to depend on the exact method of freeze-thawing. Whatever the reason, the increase following addition of serum (Fig. 11.5) is interesting. Essentially similar results were obtained with [³H] inositol- and [³H] ethanolamine-labelled cells. Analysis of [³H] choline metabolites appearing in the *medium* likewise showed the

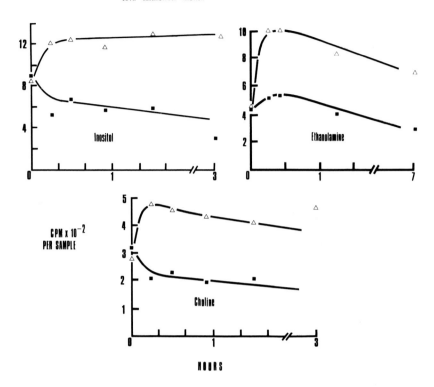

Fig. 11.4 Effect of serum on the appearance of isotope in the EDTA-extractable fraction. HTC cells were treated as described in Fig. 11.3. The cell pellet was washed by suspending in phosphate-buffered saline (pH 7.2) and centrifuging through 1% Ficoll in phosphate-buffered saline. The pellet was extracted by freeze-thawing in the presence of 0.5 mM EDTA (pH 5) and assaying radioactivity in the supernatant following centrifugation. Control, ■——■; 10% serum, △—△.

presence of labelled phosphorylcholine, some choline, and phosphatidylcholine.

The appearance of phosphatidylcholine in the EDTA extract and in the medium could be due to the "mobilization" of small pieces of membrane. Since these might also contain protein, valine-labelled cells were studied. Although no isotope appeared in the medium (Fig. 11.6) an increase in the EDTA-soluble fraction was apparent (Fig. 11.7). This was entirely due to acid-insoluble material, i.e. protein. Although it should be stressed that these are very preliminary results indeed, it does raise the possibility that the observed increase in phospholipid turnover is initiated by a prior translocation of certain membrane components. The disposition of

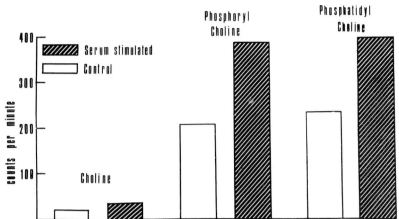

DISTRIBUTION OF METABOLITES IN EDTA-EXTRACTABLE FRACTION

Fig. 11.5 Analysis of the EDTA-extractable fraction of [³H]choline-labelled cells. The components were separated by paper chromatography (ethanol : 1 *M* ammonium acetate, pH 5.0, 7 : 3 by vol.) and the area of radioactivity cut out and assayed. Authentic choline (Rf 0.7), phosphorylcholine (Rf 0.3) and phosphatidyl-choline (Rf 1.0) were used for identification. Phospholipid at Rf 1.0 was further identified by thin layer chromatography in chloroform : methanol : acetic acid : water (50 : 30 : 25 : 4 by vol.)

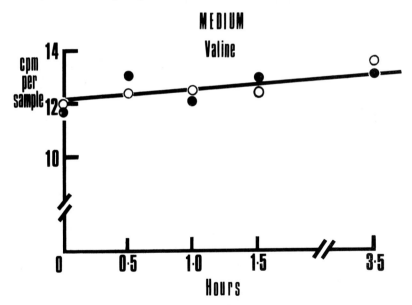

Fig. 11.6 Effect of serum on the appearance of isotope in the medium. HTC cells prelabelled with [U-¹⁴C]valine (0.07 μCi/ml) were treated as described in Fig. 11.3. Control, O—O; 10% serum, ●—●.

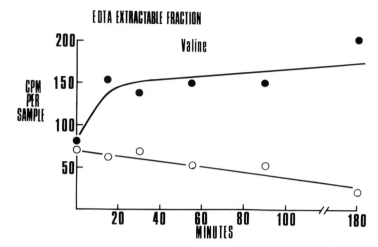

Fig. 11.7 Effect of serum on the appearance of isotope in the EDTA-extractable fraction. HTC cells prelabelled with [U-^{14}C]valine (0.07 μCi/ml) were treated as described in Fig. 11.4 Control, O—O; 10% serum, ●—●.

specific plasma membrane markers during the serum- and phyto-haemagglutinin-stimulated resumption of cellular activity is currently under study.

Acknowledgements
    We are grateful to the Christopher Welch Trustees (P.K.) and the Medical Research Council for financial support.

REFERENCES

Bergeron, J. J. M., Warmsley, A. M. H. and Pasternak, C. A. (1970) *Biochem. J. 119*, 489–492.

Cunningham, D. D. (1972) *J. Biol. Chem. 247*, 2464–2470.

Cunningham, D. D. and Pardee, A. B. (1969) *Proc. Nat. Acad. Sci. U.S.A. 64*, 1049–1056.

Cunningham, D. D. and Pardee, A. B. (1971) in *Growth Control in Cell Cultures* (Wolstenholme, G. E. and Knight, J., eds.) pp. 207–220. Churchill Livingstone, Edinburgh.

Dawson, R. M. C. (1966) in *Essays in Biochemistry* (Campbell, P. N. and Greville, G. D., eds.) Vol. 2, pp. 69–115, Academic Press, London.

Dulbecco, R. (1971) in *Growth Control in Cell Cultures* (Wolstenholme, G. E. and Knight, J., eds.) pp. 71–87 Churchill Livingstone, Edinburgh.

Fisher, D. B. and Mueller, G. C. (1968) *Proc. Nat. Acad. Sci. U.S.A. 60*, 1396–1402.

Fisher, D. B. and Mueller, G. C. (1969) *Biochim. Biophys. Acta 176*, 316–323.

Ling, N. R. (1968) *Lymphocyte stimulation*, pp. 233–264, North-Holland Publishing Co., Amsterdam.

Pasternak, C. A. (1972) *J. Cell Biol. 53*, 231–234.

Pasternak, C. A. (1973a) *Biochem. Soc. Trans. 1*, 333–336.

Pasternak, C. A. (1973b) in *Companion to Biochemistry* (Bull, A. T., Lagnado, J. R., Thomas, J. O. and Tipton, K. F., eds.) Longman, London.

Pasternak, C. A. (1973c) in *Cancer Lipids : Biochemistry and Metabolism* (Wood, R., ed.) American Oil Chemists Society. In press.

Pasternak, C. A. and Bergeron, J. J. M. (1970) *Biochem. J. 119*, 473–480.

Pasternak, C. A. and Friedrichs, B. (1970) *Biochem. J. 119*, 481–488.

Pasternak, C. A. and Micklem, K. J. (1973) In press.

Pasternak, C. A. and Quinn, P. J. (1973) *Biochim. Biophys. Acta* Biomembrane Reviews. In press.

Plagemann, P. G. W. (1971) *J. Lipid Res. 12*, 715–724.

Sefton, B. M. and Rubin, H. (1971) *Proc. Nat. Acad. Sci. U.S.A. 68*, 3154–3157.

Stoker, M. G. P. (1972) *Proc. Roy. Soc. Lond. B. 181*, 1–17.

Tomkins, G. M., Thompson, E. B., Hayashi, S., Gelehrter, T., Granner, D. and Petenkofsky, B. (1966). *Cold Spring Harbour Symp. 31*, 349–360.

Vaheri, A., Ruoslahti, E. and Nordling, S. (1972) *Nature (London)*. In press.

Warmsley, A. M. H. and Pasternak, C. A. (1970) *Biochem. J. 119*, 493–499.

Wiebel, F. and Baserga, R. (1969) *J. Cell Physiol. 74*, 191–202.

III The Viral Envelope

12 *Viral Lipids—Host Cell Biosynthetic Parameters*

H. A. Blough, W. R. Gallaher and D. B. Weinstein

Scheie Eye Institute, Division of Biochemical Virology and Membrane Research, University of Pennsylvania School of Medicine, Philadelphia, Pennsylvania

Lipids, in general, make up 20–35% by weight of most animal viruses (Blough and Tiffany, 1971). There are of course exceptions, i.e. vaccinia virus contains but 5% lipid (Zwartouw, 1964). Lipids, when interacted with protein, make up the viral envelope—that anatomical part of the virion which serves as a molecular container for the nucleocapsid. While it has been tacitly assumed, in the past, that lipids are incorporated passively into the envelope and the *de novo* synthesis does not occur during infection, recent studies have, in fact, shown that several factors determine the amount and species of the lipid incorporated into the virion. These are (1) an "envelope" structural polypeptide(s) which selects out appropriate acyl chains due to the primary amino acid sequence (Blough and Lawson, 1968; Tiffany and Blough, 1969a); (2) environmental factors (Blough *et al.*, 1967; Blough and Tiffany, 1969); (3) alterations in host cell biosynthetic and catabolic pathways (Blough and Lawson, 1968).

This chapter deals with a comparative study of the effect on host cell lipid metabolism by two RNA-containing viruses which acquire their envelopes at the cell surface membrane, i.e. influenza (Hoyle, 1950) and Newcastle disease viruses, as well as by Herpes virus, an icosahedral DNA-containing virus which is preferentially enveloped at the inner nuclear membrane (Siegert and Falke, 1966). This type of study will allow us to compare the effect of different virus infections on lipid metabolism from the time of virus absorption to cell lysis and/or fusion.

In general, unless otherwise stated, cells (BHK-21, fibroblasts etc.) were grown to confluence in minimal Eagle's medium or Puck's medium containing 10% calf serum. Cells were then starved of lipid for 12–18 h in a chemically-defined medium containing 0.1–1.0% fatty acid-poor bovine serum albumin (in place of serum). Virus was adsorbed at 4° or 23° (the low temperature serves a dual purpose: it allows virus absorption but not penetration to occur and at the same time shuts off host cell macromolecular synthesis). Cells were then pulsed at appropriate time intervals with 2-[^{3}H]-glycerol and [^{14}C]-acetate at 37° and chased in carrier-free medium. Monolayers were washed, scraped off the glass and the lipids were extracted with $CHCl_3$-CH_3OH (2:1, v/v) five times. Neutral lipids were separated on 36 cm plates using the system of Freeman and West (1966). Phospholipids were separated by two dimensional thin-layer chromatography (TLC) on plates of silica gel G or H (Blough et al., 1967). Samples were quantified by liquid scintillation spectroscopy and these values were converted to specific activities by assaying for inorganic phosphorus or by charring.

EXPERIMENTAL ISOTOPIC PROCEDURES

In order to study lipid synthesis, 2-[^{3}H]glycerol was used to label the glycerol backbone and [^{14}C]-acetate the sterols, free fatty

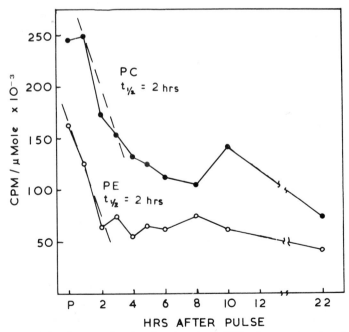

Fig. 12.1   Pulse-chase studies showing that the half life of two major phospholipids is ca. 2 h.

acids and fatty acyl chains. Under the conditions of our experiments, 95% of the glycerol label goes to the glycerol skeleton and 85% of the acetate label to the fatty acyl chains of glycerides and phospholipids. Short pulses were used since minor changes would be missed with longer pulses. It has been shown by Pasternak and Bergeron (1970) and Plagemann (1971) that the phospholipids turn over slowly, viz. these molecules possess a half life of 15–24 h. In order to assess this important feature, we pulsed cells for one hour with 2-[^{3}H]glycerol containing 1–3 μCi/ml and then chased for various time intervals. Lipids were quantified as described above. As seen in Fig. 12.1, the specific activity of phosphatidylcholine and phosphatidylethanolamine (which comprise 75% of total phospholipids of BHK 21 cells) have a half life of 2.0–2.5 h. These findings were reproducible for both confluent and contact-inhibited cultures and could be duplicated with 2-[^{14}C]glycerol in CH$_3$[^{3}H]choline.

VIRUS INFECTED CELLS

A. Chick Embryo Fibroblasts Infected by Orthomyxoviruses

We observed that under conditions of low multiplicities of infection (3–5 pfu*/cell) the level of *de novo* synthesis of host cell phospholipids of chick fibroblasts continued unchanged until seven hours post infection with A$_0$/WSN strain of influenza virus. At seven hours, phosphatidic acid, sphingomyelin and phosphatidylinositol were depressed (Fig. 12.2). At 13 h all phospholipids were depressed 56–94%. Furthermore when we increased the multiplicity of infection to 20 pfu*/cell, the lipid synthesis was depressed as early as 4 h (Fig. 12.3). The finding of continued synthesis of lipid following infection disproved the conclusion that *de novo* synthesis of lipid did not occur in virus-infected cells (Kates *et al.*, 1961) and stresses the importance of using multiple isotopic labelling techniques.

Turnover studies were less conclusive. In the chick embryo fibroblast system, although there was an overall depression of lipid synthesis 8 h after infection; turnover of the glycerol backbone (during a four-hour chase) appears enhanced in infected cells in the case of phosphatidylcholine and phosphatidylserine. The fact that turnover rates appear remarkably similar in controls and infected cells despite an 80–90% decrease in phospholipid synthesis at these times is a paradox. Possible explanations are that preformed phospholipid molecules are transported to the cell surface membranes for assembly in viral envelopes. Alternatively, that portion of lipid

* Pfu: plaque forming units.

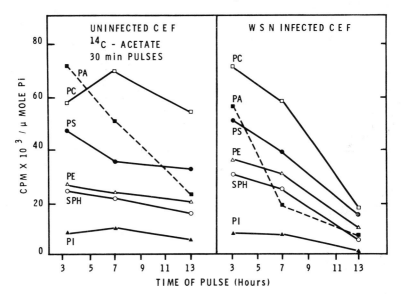

Fig. 12.2   Effect of influenza virus (A_0/WSN) on phospholipid synthesis in chick fibroblasts using a double label of 2-[^{3}H]glycerol and [^{14}C]acetate. Cells were pulsed for 30 min at various time intervals after infection and sampled immediately. Depression of synthesis of various phospholipid species is seen to occur 7 h after infection.

which is turning over very rapidly is necessary for viral envelope biogenesis.

It has been shown by Lands (1965) that enzymes of the mono-acyl-diacyl phosphoglyceride cycle are responsible for the renewal of fatty acyl chains in a variety of tissues. These enzymes are located primarily in the microsomal and mitochondrial fractions of cells (Stoffel and Scheifer, 1968; Eibl *et al.*, 1970). Due to the varying acyl chain composition observed in different strains of myxoviruses grown in embryonated eggs (Tiffany and Blough, 1969a, b), this system was investigated to see whether influenza or Newcastle disease viruses could alter host cell enzymes responsible for phospho-lipid renewal. Microsomal fractions of chorioallantoic cells were isolated at various times post infection, and compared with uninfected controls. As early as 30 min post infection, it was seen that the phosphoglyceride acyl-CoA transferase was depressed ca. 70% (Fig. 12.4). The acyl-CoA synthetase was unaffected and the acyl CoA hydrolase was enhanced threefold with incomplete influenza virus of the von Magnus type at 40 h post-infection (not shown). These enzymes were *not* detected within the virion. The conclusion of the studies was that, *in ovo*, the enzymes of the monoacyl-diacyl

PHOSPHOLIPID SYNTHESIS IN INFLUENZA VIRUS
INFECTED CELLS (20 PFU/CELL)

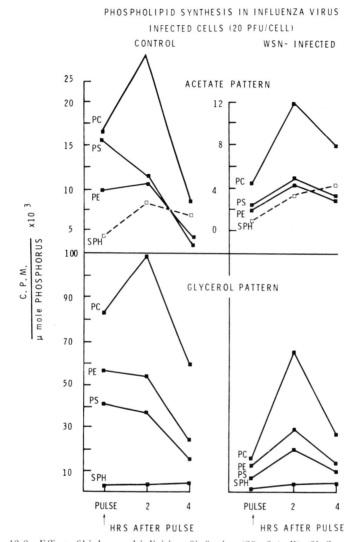

Fig. 12.3 Effect of higher multiplicities of infection (20 pfu/cell) of influenza virus on host cell lipid metabolism. Cells pulsed at four hours post infection. Note immediate depression of all phospholipids.

phosphoglyceride cycle did not play a major role in determining the molecular species of phospholipid incorporated into the viral envelope (Blough and Smith, 1973).

Experiments were also designed to try to test the hypothesis that altered lipid metabolism in cells infected with certain viruses (e.g. cells infected with incomplete influenza virus) or certain non-permissive cell types (e.g. HeLa cells) are responsible, in part, for

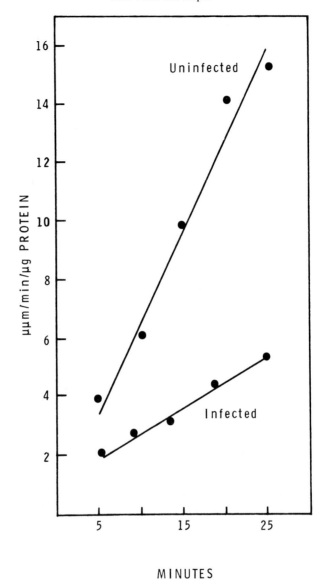

MINUTES

Fig. 12.4 Inhibition of host cell acyl-CoA phosphoglyceride acyl transferase activity following infection with influenza virus. The reaction mixture contained 20 mμM of lysophosphatidylcholine as the acceptor, 200 μg of microsomal protein from chick chorioallantoic cells, and 5.5′ dithio-*bis*-nitrobenzoic acid (DNTB); the molar adsorptivity was used to calculate the release of free mercaptan. Microsomes were isolated 30 min after infection.

failure of viral assembly. Influenza virus which has been passed at
high multiplicity becomes pleomorphic and loses a portion of its
genome (Pons and Hirst, 1969). This behaves in fact like a deletion
mutation—the so-called incomplete or von Magnus virus. When
compared to standard virus, von Magnus virus has an altered lipid
composition and shows obvious defects in the viral envelope (Blough
and Merlie, 1970). When incorporation studies were done there was
an immediate 50% decrease in diglyceride synthesis and a 75%
increase in free fatty acids when compared to cells mock-infected
with sterile allantoic fluid (Table 12.1). Furthermore, we observed

Table 12.1

EFFECT OF INCOMPLETE VIRUS ON INCORPORATION OF
^{3}H-GLYCEROL AND ^{14}C-ACETATE INTO NEUTRAL LIPIDS OF
CHICK EMBRYO FIBROBLASTS*

| | ^{3}H-Glycerol | | | ^{14}C-Acetate | | |
	Control	Infect.	% Change	Control	Infect.	% Change
Monoglyceride	40,029	51,821	+29	44,000	46,607	+6
Diglyceride	613,760	521,503	−15	272,552	135,134	−50
Triglyceride	167,193	186,184	+11	14,570	15,225	+5
Fatty Acids	—	—	—	10,175	17,809	+75
Cholesterol	—	—	—	79,351	64,210	−19
Cholesterol Esters	—	—	—	766	762	NC

* Specific Activity: CPM/μM of neutral lipid
Pulse immediately after virus adsorption.

an enhanced synthesis of phosphatidylcholine and phosphatidylethanolamine at 6 h in chick embryo cells infected with von Magnus
virus (Fig. 12.5). These results correlate well with compositional
analysis of von Magnus virus (Blough and Merlie, 1970) in which
free fatty acids and phosphatidylethanolamine were increased in
comparison to standard virus. Obviously the presence of these
charged molecules within the envelope would render it unstable.

B. *Changes in Lipid Synthesis During Fusion From Without Induced by
Newcastle Disease Virus (NDV)*

The molecular mechanism of cell fusion is unknown. However,
it appears that fusion occurs by the dissolution or rearrangement of
existing cell membranes. By studying alterations in lipid synthesis or
composition during induction of cell fusion by NDV it was hoped
that we could correlate these events with biochemical changes in
host cell lipids. Instead we found that lipid synthesis is generally

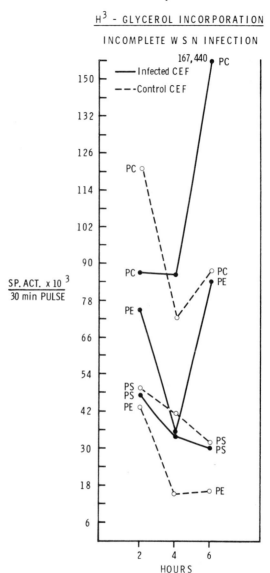

H^3 - GLYCEROL INCORPORATION

INCOMPLETE W S N INFECTION

Fig. 12.5 Stimulation of phosphatidylcholine and phosphatidylethanolamine synthesis at 6 h post infection with von Magnus virus.

suppressed during cell fusion and that lipid composition of fusing cells *in vitro* remains quite stable.

The cells used in our study were monolayer cultures of BHK-21 cells or secondary chick embryo cells, plated in a maintenance medium containing 2% foetal calf serum, at a density where most cells are touching but not yet confluent. The level of fusion used in the following experiments averaged 1.0 fusion event* per cell. The type of fusion has been called fusion from without (FFWO) by Bratt and Gallaher (1969), and corresponds to the Sendai-virus-induced fusion used in cell hybridization studies.

Our previous protocol was modified for these experiments. The medium was removed and cells washed twice with minimal Eagle's medium (MEM) to remove serum components. The plates were then cooled and 10,000 Haemagglutinating units/ml (HAU/ml) of the N.J. La Sota strain of NDV (NDV-N) were adsorbed for 1 h at 4°. After removal of the inoculum, 1–3 μCi/ml of 2-[^{3}H]-glycerol was added (at 38°) and the cultures pulsed for 30 min. The label was then removed, cells washed twice, and fresh medium added for chase periods of 30 to 90 min for maximal labelling of phospholipid, at which times cultures were harvested and lipids extracted.

The results of these experiments are shown in Table 12.2, in which the specific activities of phospholipids from NDV-treated

Table 12.2
EFFECT OF NEWCASTLE DISEASE VIRUS (NDV)-TREATMENT
ON LIPID SYNTHESIS IN CHICK EMBRYO CELLS

Lipid component	NDV-N treated as fraction of control specific activity		
	PULSE	30 min chase	90 min chase
Phosphatidylserine	0.72	0.48	0.68
Phosphatidylcholine	0.71	0.28	0.48
Phosphatidylethanolamine	0.75	0.50	0.81
Phosphatidylinositol	1.67	0.29	0.32
Total phospholipid	0.77	0.32	0.47

cells are expressed at a fraction of the controls. Immediately following the pulse there was a 23% overall suppression of lipid synthesis, and 53–68% suppression of maximal incorporation of label during the chases. The degree of suppression depends on the species of phospholipids. Thus, the syntheses of phosphatidylserine and phosphatidylethanolamine were more resistant to inhibition in chick embryo cells than phosphatidylcholine, sphingomyelin or phosphatidylinositol. Similar results were obtained, in both the labelling pattern and kinetics, in BHK-21 cells.

In an attempt to determine how drastic a suppression of lipid synthesis this represented physiologically, lipid synthesis was

* Fusion events: average number of nuclei per uninfected cell subtracted from the average number of nuclei per cell of infected cultures.

measured in cells lysed by NDV-N in the absence of Ca^{++} and also in cells infected with a virulent strain of NDV which induced cell fusion after a productive infection. In general, the suppression of lipid synthesis seen during FFWO was ca. one-half that observed with cytolysis, and greater than that observed until very late in infection.

We determined whether this suppression of lipid synthesis resulted in a net breakdown of phospholipid or change in lipid composition. Of specific interest was the possible generation of lysocompounds such as lysophosphatidylcholine, which have been shown by Poole *et al.* (1970) to induce fusion. In comparing control and NDV-N treated BHK-21 cells the mole % of total lipid phosphorus of each major phospholipid, as well as those areas where lysocompounds chromatograph, were determined. There were little differences between control and NDV-N treated cells (Table 12.3).

Table 12.3
EFFECT OF NEWCASTLE DISEASE VIRUS (NDV)-TREATMENT
AND CELL FUSION ON THE LIPID COMPOSITION
OF BHK-21 CELLS

Lipid Component	Mole % of Total Lipid Phosphorus	
	Control	NDV-N treated
Origin + Lysophosphatidylcholine	4.4	5.4
Sphingomyelin	10.0	6.5
Phosphatidylserine	13.5	10.2
Phosphatidylcholine	38.2	35.4
Phosphatidylethanolamine	25.3	27.9
Phosphatidylinositol	2.7	2.7

Finally we wished to determine whether the suppression of lipid synthesis we observed was due specifically to cell fusion, or whether there was a general effect of high multiplicity of infection as occurred with influenza viruses. It had been shown that the level of fusion can be varied in infected cells by incubating these cells for various periods at room temperatures (Gallaher and Bratt, 1972). If, following virus adsorption at 4° cultures are shifted up to 38° and held at 60 min, 1.27 fusion events per cell were observed. If cultures were treated with the same concentration of virus, but initially incubated at 23° for 30 min before shifting to 38°, these cultures undergo only 0.24 fusion events per cell. By comparing temperature shifts on fusion with the effect on incorporation of 2-[^{3}H]glycerol the results shown in Fig. 12.6 are obtained. The per cent of maximum fusion and per cent of control phospholipid

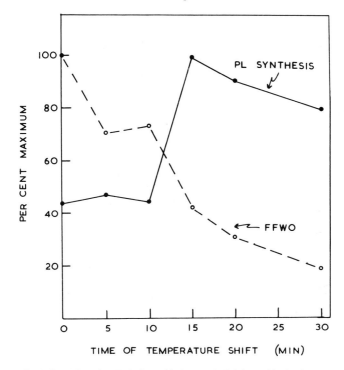

Fig. 12.6 Relationship of cell fusion, high multiplicities of infection and phospholipid synthesis in BHK-21 cells infected with Newcastle disease. (Fusion from without.) Temperature step-up experiments.

synthesis was co-plotted against the time of temperature shift from from 23° to 38°. As fusion was inhibited the incorporation of glycerol into phospholipid approached those of the control. Since NDV-N at a concentration of 10,000 HAU/ml does not always suppress lipid synthesis, but only under conditions when fusion was induced, it appeared that the inhibition of lipid synthesis we observed with NDV-N was not simply a general effect of high multiplicity of infection.

C. *Herpes Virus Infected BHK-21 Cells*

Studies were done in our laboratory on BHK-21 cells which were infected with Herpes simplex virus using short pulse chase experiments done as described above. Herpes virus does not fuse cells from without but causes fusion late in infection. At 3 to 6 h, all lipids remain relatively stable (Fig. 12.7a). At 9 h, there was a stimulation of both isotopic precursors flowing into all phospholipids (Weinstein and Blough, 1972); this is the time of cell fusion.

Table 12.4
LIPID COMPOSITION OF HERPES-INFECTED CELLS

| | | MOLE PER CENT | | | | |
| LIPID | CONTROL RANGE | TIME AFTER INFECTION | | | | |
		3 h	6 h	9 h	12 h	15 h
Monoglycerides	16.0–17.4	16.2	14.1	17.8	—	—
Diglyceries	19.5–22.9	22.9	20.2	22.5	—	—
Triglycerides	8.6–11.6	8.8	10.5	8.6	—	—
Cholesterol	25.4–27.9	27.9	27.7	25.4	—	—
PC	43.0–46.3	43.8	42.1	45.3	46.6	44.5
PS	10.5–11.9	12.3	13.5	11.3	7.9	10.5
PE	26.2–31.7	30.5	25.6	27.7	36.0	26.9
SPH	5.0–11.7	4.7	11.1	7.4	4.7	9.8

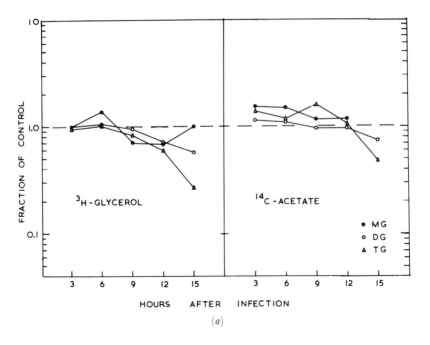

(a)

Fig. 12.7a   Utilization of [³H]glycerol (2 μCi/ml) and [¹⁴C]acetate (0.5 μCi/ml) of uninfected BHK-21 cells and these BHK-21 cells infected with Herpes simplex virus. Individual neutral lipids were separated on 36 cm plates of silica gel G, and specific activities determined. Samples were taken immediately following a 30 min pulse, and at intervals during a 4 h chase period. (MG = monoglycerides; DG = diglycerides; TG = triglycerides).

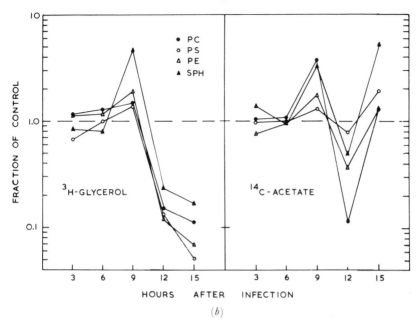

Fig. 12.7b   Cyclic changes in phospholipid synthesis when plotted as a fraction of uninfected controls. Note apparent stimulation of acetate incorporation late on in infection (when glycerol flow into phospholipids is depressed).

At 12–15 h, synthesis of all phospholipids were depressed (Fig. 12.7b). The different rates of uptake of acetate and glycerol into the same lipids late in infection suggest different pools may be operative late in the infectious cycle. Despite the cyclic changes in phospholipids in Herpes virus infected cells, the composition of host-cell lipids during the infectious cycle was stable, and increased levels of lysophosphatides were not detected (Table 12.4). This suggested that monoacylated phospholipids *do not* play a part in virus induced cell fusion. Turnover of diglycerides and triglycerides (Fig. 12.8) were the same in infected and uninfected cells, *viz*, $2–2\frac{1}{2}$ h half-lives for both molecules (Blough, 1971). Fatty acids, sterols and monoglycerides did not turn over under these experimental conditions. With phospholipids, the glycerol skeleton shows a differential pattern of turnover of phosphatidylserine and phosphatidylcholine when comparing infected to uninfected cells.

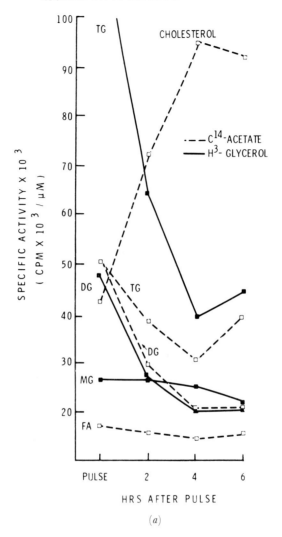

NEUTRAL LIPID METABOLISM IN BHK-21 CELLS.

Fig. 12.8a and *b*   Synthesis and turnover of neutral lipids at 4 h post-infection with Herpes simplex virus in BHK-21 cells.

These cyclic changes are similar to those described by Ben-Porat and Kaplan (1971) for some species of phospholipid. It may well be that definitive measurements of turnover must await the isolation and quantitation of isotopic precursor and cellular product pool sizes before we state empirically that turnover is affected. Obviously there may be a great deal of variation of lipid synthesis and turnover

HERPES SIMPLEX INFECTED BHK-21 CELLS.

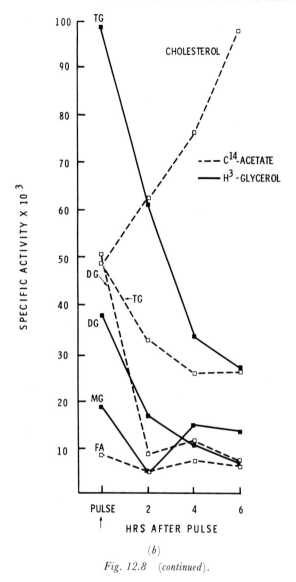

(b)

Fig. 12.8 (continued).

in cultured cells, which no doubt reflects cell types (multiple in the case of primary chick embryo fibroblasts), cell cycle conditions, contact inhibition and other physiological and metabolic events responsible for cell growth and division.

In conclusion, several points are to be stressed. In uninfected cells, many phospholipids and di- and triglycerides turn over very rapidly *viz* with a half-life of $2-2\frac{1}{2}$ h. In general, infection by enveloped viruses is accompanied by a suppression of lipid synthesis, as shown for Sindbis virus by Pfefferkorn and Hunter (1963) and the time of the depression is dependent upon the multiplicity of infection. Exceptions to this are Herpes-virus and von Magnus virus which cause a stimulation of all or some classes. The glycerides are more stable than phospholipids and are affected late in the infectious cycle. Although fusion from without (with NDV) suppresses lipid synthesis, fusion from within (by Herpes) appears to enhance it. Despite the marked rearrangement of membrane components associated with fusion, the lipid composition of the host cell remains remarkably stable.

Acknowledgements
This work was supported by the Commission on Influenza, Armed Forces Epidimiological Board, through the U.S. Army Medical Research and Development Command, Department of the Army (Research Contract no. DADA 17-67C-7128). HAB is a Senior Post-Doctoral Fellow of the National Multiple Sclerosis Society; WRG is a Post-Doctoral Fellow of the National Cancer Institute (1FO2-CA-50429-1). DBW was a Post-Doctoral Fellow of the National Institute of Allergy and Infectious Diseases (1-FO2-AI-48,867-01) and the Fight for Sight, Inc., New York City.

REFERENCES

Ben-Porat, T. and Kaplan, A. S. (1971) *Virology 45*, 252.
Blough, H. A., Weinstein, D. B., Lawson, D. E. M. and Kodicek, E. (1967) *Virology 33*, 459.
Blough, H. A. and Lawson, D. E. M. (1968) *Virology 36*, 286.
Blough, H. A. and Tiffany, J. M. (1969) *Proc. Natl. Acad. Sci. U.S.A.*, *62*, 242.
Blough, H. A. and Merlie, J. P. (1970) *Virology 40*, 685.
Blough, H. A. (1971) *Proc. 2nd Intl. Congress Virology Budapest* P. 133.
Blough, H. A. and Tiffany, J. M. (1973) *Advanc. Lipid Res.*, *11*, 267, in the press.
Blough, H. A. and Smith, W. R. (1973) *J. gen. Virology* in the press.
Bratt, M. A. and Gallaher, W. R. (1969) *Proc. Natl. Acad. Sci. U.S.A. 64*, 536.
Eibl, H., Hill, E. E. and Lands, W. E. M. (1970) *European J. Biochem.*, *9*, 250.
Freeman, C. P. and West, D. (1966) *J. Lipid Res.*, *7*, 324.
Gallaher, W. R. and Bratt, M. A. (1972) *J. Virol.*, *10*, 159.
Hoyle, L. (1950) *J. Hyg. London 48*, 277.
Kates, M., Allison, A. C., Tyrrell, D. A. J. and James, A. T. (1961) *Biochim. Biophys. Acta 52*, 455.
Lands, W. E. M. (1965) *Advanc. Biochem. 34*, 313.
Pasternak, C. A. and Bergeron, J. J. M. (1970) *Biochem. J. 119*, 473.
Pfefferkorn, E. R. and Hunter, H. S. (1963) *Virology 20*, 446.
Plagemann, P. G. W. (1971) *J. Lipid Res. 12*, 715.
Pons, M. and Hirst, G. (1969) *Virology 38*, 68.
Poole, A. R., Howell, J. I. and Lucy, J. A. (1970) *Nature London 227*, 810.

Siegert, R. S. and Falke, D. (1966) *Arch. Virusforsch. 19*, 230.
Stoffel, W. and Scheifer, H.-G. (1968) *Z. physikal. Chem. 349*, 1017.
Tiffany, J. M. and Blough, H. A. (1969a) *Science 163*, 573.
Tiffany, J. M. and Blough, H. A. (1969b) *Virology 37*, 492.
Zwartoun (1964) *J. Gen. Microbiol., 34*, 115.

13 *Glycoproteins and Glycolipids in Viral Envelopes*

Hans-Dieter Klenk

Institut für Virologie, der Justus Liebig-Universität, Giessen, Germany

Many animal viruses such as togaviruses, rhabdoviruses, myxoviruses, and the RNA-tumour viruses are assembled and released by budding from the surface of the host cell. During this process the internal component of these viruses is enveloped by a membrane which is continuous with the plasma membrane of the host cell (Compans *et al.*, 1966; Compans and Dimmock, 1969; Bächi *et al.*, 1969). Viral envelopes meet all the criteria for cellular membranes. Morphologically they possess a trilaminar or unit membrane structure. Chemically they are composed of proteins, lipids, and carbohydrates (Klenk and Choppin, 1969a). The lipids of the viral envelope are derived from the plasma membrane of the host cell (Klenk and Choppin, 1969b). There are only very few proteins in the envelope. They are all virus-specific (Choppin *et al.*, 1971). The simplicity of the protein pattern greatly facilitates studies on envelope structure and assembly and on the interactions of proteins and lipids. Therefore, viral envelopes represent valuable tools for the investigation of membrane structure and biogenesis in general.

Although the presence of carbohydrates in viruses has been known for some time, it has not been until recently that more detailed information about their structure and biosynthesis has been available and there is still very little knowledge about the biological significance of these envelope constituents.

Most of the data presented in this article have been obtained from experiments with influenza virus and the parainfluenza virus SV5 both of which belong to the myxovirus group. There is evidence that the basic principles derived from these data are also valid for other virus groups such as toga-, rhabdo- and RNA-tumourviruses (*cf.* Klenk, 1973).

THE STRUCTURE OF THE ENVELOPE

The influenza virus used for these investigations is fowl plague virus (FPV) which belongs to the influenza A group. The polypeptides of the virus grown in chick embryo fibroblasts have been determined by polyacrylamide gel electrophoresis (Klenk *et al.*, 1972a). Six polypeptides have been found in the virion with molecular weights ranging from 26,000 to 85,000 (Figs. 13.1 and 13.2). Polypeptides 1, 2 and 6 are carbohydrate-free; polypeptides 3, 4 and 5 are glycoproteins.

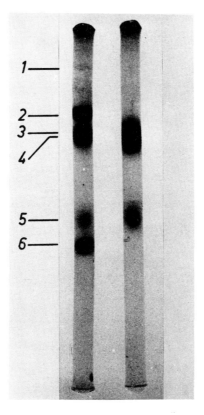

Fig. 13.1 Polypeptides of fowl plague virus separated by polyacrylamide gel electrophoresis and stained with Coomassie blue. Intact virions on the left and isolated spikes on the right. From Klenk *et al.* (1972a).

These polypeptides have been localized within the virion by stepwise degradation of the virus and by analysis of the resulting subviral particles (Klenk *et al.*, 1972a). Electron microscopy reveals that influenza virus like most enveloped viruses has characteristic surface projections, the so-called spikes. These spikes can be split from the virion and isolated in pure form. Such preparations which contain haemagglutinin and neuraminidase, biological properties characteristic for myxoviruses, consist exclusively of the three glycoproteins (Fig. 13.1). The observation that the viral glycoproteins are located in the spikes has been made not only with influenza virus but also with a whole series of other enveloped

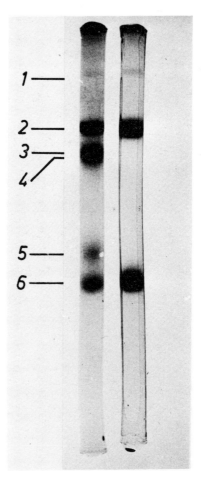

Fig. 13.2 Polypeptides of fowl plague virus separated by polyacrylamide gel electrophoresis and stained with Coomassie blue. Intact virions on the left and spike-free particles on the right. From Klenk *et al.* (1972a).

viruses (Compans *et al.*, 1970; Chen *et al.*, 1971; McSharry *et al.*, 1971; Rifkin and Compans, 1971) which suggests that this is a structural feature common to all viral envelopes. In this respect viral envelopes resemble also cell membranes which have glycoproteins on their surface, too (Uhlenbruck, 1971).

Treatment with proteolytic enzymes removes the spikes from the virion (Compans *et al.*, 1970; Schulze, 1970). The spike-free particles have lost all glycoproteins; the carbohydrate-free polypeptides and the entire lipid including the glycolipids remain in these particles (Fig. 13.2).

These and other data (Compans *et al.*, 1972; Schulze, 1972; Klenk *et al.*, 1972a) suggest that the viral envelope consists of a central lipid layer which is surrounded on its outer surface by a layer of glycoprotein spikes, whereas its inner surface is coated by a carbohydrate-free polypeptide (Fig. 13.3).

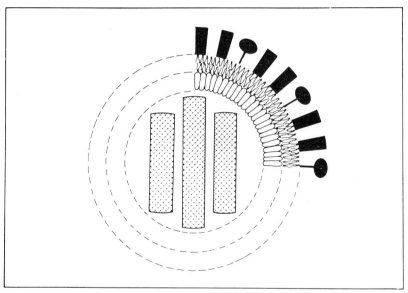

Fig. 13.3   A model of the structure of the envelope of influenza virus. The lipid is arranged in a bilayer at a radius of about 35 nm, and the spikes are attached to the outer surface of the bilayer but do not penetrate it. The rod-like shape of the haemagglutinin spikes and the knob-like shape of the neuraminidase spikes are adapted from Laver and Valentine (1969). On the inner surface of the lipid bilayer is a continuous layer of protein subunits of about 6 nm thick, and below this layer is the nucleocapsid. From Compans *et al.* (1972).

BIOGENESIS OF THE HAEMAGGLUTININ SPIKES

Data reported by several investigators indicate that the haemagglutinin of influenza virus is composed of two glycoprotein molecules of the molecular weight of about 50,000 (HA$_1$ in Fig. 13.4) and two glycoprotein molecules of the molecular weight of about 30,000 (HA$_2$ in Fig. 13.4) (Skehel and Schild, 1971; Laver, 1971; Stanley and Haslam, 1971). These form the functional haemagglutinating unit (MW ca. 150,000) which corresponds to a spike.

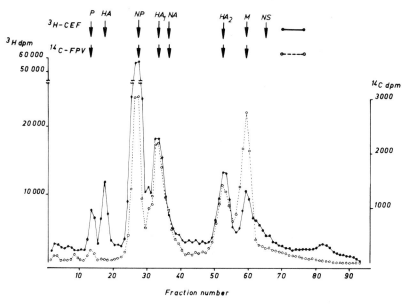

Fig. 13.4 Coelectrophoresis in a polyacrylamide gel of the proteins from [^{3}H] amino acid-labelled infected chick embryo fibroblasts with the proteins of [^{14}C] amino acid-labelled purified fowl plague virus. Cells were labelled 4 h post infection for 1 h. Arrows indicate the position of polypeptides. The relationship between the nomenclature in this figure and that in Figs. 13.1 and 13.2 is as follows: P = 1, NP = 2, HA$_1$ = 3, NA = 4, HA$_2$ = 5, M = 6. From Klenk *et al.* (1972b).

Studies of the biosynthesis of influenza proteins are greatly facilitated by the fact that the synthesis of host macromolecules is suppressed in infected cells. Thus, in pulse-label experiments with radioactive amino acids only virus-specific proteins are labelled. Such experiments revealed in FPV-infected cells the presence of all structural proteins of the virion and in addition two non-structural proteins (Fig. 13.4) (Klenk *et al.*, 1972b). The non-structural protein HA (MW 76,000) is a glycoprotein and a precursor of the structural glycoproteins HA$_1$ and HA$_2$. This has been demonstrated by pulse-chase experiments (Lazarowitz *et al.*, 1971; Klenk *et al.*, 1972a) and it has been confirmed by experiments in which virus-infected cells have been kept at 25°. Under these conditions glycoprotein HA accumulates in the cell, whereas glycoproteins HA$_1$ and HA$_2$ are missing (Fig. 13.5). However, when the temperature is shifted to 37°, HA decreases drastically and HA_1 and HA_2 appear.

By blocking the haemagglutinin biosynthesis at a different stage we have been able to detect another precursor protein (Klenk *et al.*, 1972b). The inhibitor employed was D-glucosamine which in

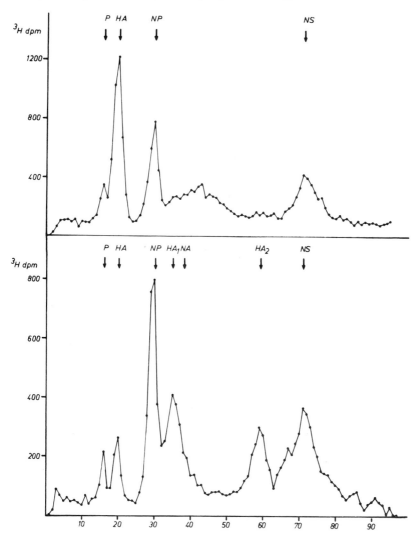

Fig.13.5   Upper panel: Electrophoresis of fowl plague virus proteins synthesized in chick embryo fibroblasts at 25°C. Cells were labelled 24 h p. i. by a 1 h pulse with [³H]valine. Lower panel: Infected cells were kept at 25°C for 24 h and labelled as described above. The radioactive pulse was then followed by a chase with non-radioactive valine at 37°C for 3 h. Labelling conditions have been described in detail previously (Klenk *et al.*, 1972b).

cultured cells is converted into its UDP-derivative. Thus high doses of glucosamine deplete the UTP-pool of the cell and therefore the pool of sugar nucleotides. If these activated sugars are missing, the glycosylation of proteins should be impossible and therefore the biosynthesis of glycoproteins inhibited. Under these conditions in FPV-infected cells glycoproteins HA, HA_1 and HA_2 indeed can no longer be detected (Fig. 13.6). However, a novel polypeptide with a molecular weight of 64,000 has been found. It has been designated HA_0, because the available evidence suggests that it is the unglycosylated or incompletely glycosylated polypeptide of glycoprotein HA.

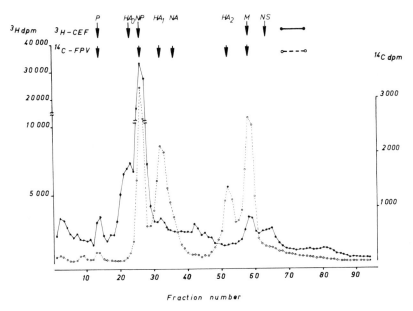

Fig. 13.6 Polyacrylamide gel electrophoresis of fowl plague virus proteins synthesized in chick embryo fibroblasts in the presence of glucosamine. Cells were labelled at 4 h p. i. for 1 h with [³H] amino acids. Marker virus labelled with [¹⁴C] amino acids was coelectrophoresed on the same gel. From Klenk *et al.* (1972b).

Thus, the first identifiable intermediate in the synthesis of the influenza virus haemagglutinin is a virus-specific polypeptide, HA_0. This polypeptide is glycosylated most probably by host-specific transferases (*cf.* Klenk, 1973). A large glycoprotein HA is formed. By cleavage of a peptide bond HA yields 2 smaller glycoproteins HA_1 and HA_2 which are both constituents of the haemagglutinin of the mature virion (Lazarowitz *et al.*, 1971).

The other functional subunit of the influenza virus envelope also located in the spike layer but clearly distinct from the haemagglutinin is the viral neuraminidase. More information is required before the structure of the neuraminidase as it exists in the virion can be determined.

It has been calculated that an influenza virion contains about 550 spikes (Tiffany and Blough, 1970; Schulze, 1972). The spikes are attached to the surface of the lipid layer and there is evidence that they do not penetrate through it (Landsberger *et al.*, 1971; Klenk *et al.*, 1972a).

LIPIDS AND GLYCOLIPIDS

The lipids are arranged in the influenza virion as a continuous bimolecular leaflet which completely separates the glycoproteins of the spikes from all other viral proteins and the nucleic acid (Fig. 13.3). This model has been derived from degradation studies with proteolytic enzymes (Klenk *et al.*, 1972a) and from electron spin resonance (Landsberger *et al.*, 1971) and x-ray diffraction (Compans *et al.*, 1972) studies.

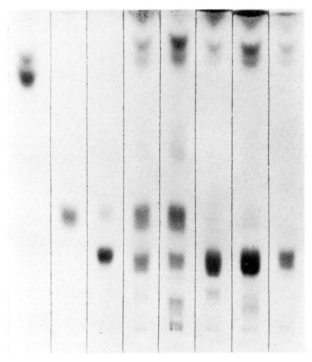

Fig. 13.7 Thin-layer chromatogram of neutral glycolipids of monkey kidney (MK) and bovine kidney (MDBK) cells, plasma membranes, and SV5 virions. (1) galactosylceramide, (2) galactosyl-galactosyl-glucosylceramide, (3) N-acetyl-galactosaminyl-galactosyl-galactosyl-glucosylceramide, (4) MK cells, (5) SV5 grown in MK cells, (6) MDBK cells, (7) MDBK cell plasma membranes, (8) SV5 grown in MDBK cells. From Klenk and Choppin (1970b).

In contrast to the virus-specific proteins the lipids in the virion are derived from the plasma membrane of the host cell. Therefore they show host specificities. This could be shown in comparative analyses of the phospholipids and the neutral lipids of the parainfluenza virus SV5 and of plasma membrane preparations of various host cells in which the virus was grown (Klenk and Choppin, 1969b, 1970a). Studies on glycolipids reveal such differences in an even more convincing manner (Klenk and Choppin, 1970b). Of particular importance is the fact that qualitative differences between various cell membranes are reflected in the virion, e.g. galactosyl-galactosyl-glucosylceramide is found in monkey kidney cells and virions grown in them, but not in membranes or virions from bovine kidney cells (Fig. 13.7). Such differences in the lipids of virions which, though grown in different cells, contain the same virus-specific proteins emphasize the importance of the host cell membrane in determining the lipid and the carbohydrate composition of the virion.

LOCALIZATION OF ENVELOPE CARBOHYDRATES BY AGGLUTININS

Glycolipids are antigenic. Nevertheless, their presence in the intact virion cannot be detected by serological methods. For instance, antisera against blood group A substance do not react with intact virions. However, when the spikes are removed from the particle, the blood group-specific glycolipids are unmasked and become serologically active. On the same line, an avertebrate agglutinin from *Helix pomatia* and a phytoagglutinin from *Dolichos biflorus* which are both specific for the terminal N-acetylgalactosamine residue of blood group A substance react only with stripped virions. Since the glycoproteins have been completely removed, the reaction can involve only the lipid-bound carbohydrates which are still present in these particles, as shown by chemical analysis (Klenk et al., 1972a).

On the other hand, a phytagglutinin of different specificity, Concanavalin A, precipitates only intact virions. If the spikes have been removed, the reaction is negative. This shows that only the spikes contain the Concanavalin A—specific receptor. The sugar determinant for this type of interaction is most likely mannose, which has been found only in viral glycoproteins (Klenk et al., 1970a), not in viral glycolipids (Klenk and Choppin, 1970b).

These data indicate that the biological specificities of the lipid- and protein-bound carbohydrates are different. The carbohydrate groups of the glycolipids are hidden in deeper layers of the envelope, in intact virions they are not exposed to the surface. They are accessible for macromolecules such as phytagglutinins and antibodies only after removal of the spikes.

Not only virus particles can be agglutinated by Concanavalin A, but also cells infected with a whole series of enveloped viruses (Becht *et al.*, 1972). This finding is interesting, because so far it has been generally assumed that this reaction was quite specific for transformed cells (Burger, 1969; Inbar and Sachs, 1969). Our results suggest that Concanavalin A induces agglutination by bridging viral spikes which appear at the surface of infected cells and carry the determinant sugar in their carbohydrate moiety.

THE CARBOHYDRATE COMPOSITION OF MYXO-VIRUS ENVELOPES

Detailed carbohydrate analyses have been carried out on several viruses which mature by budding from plasma membranes (*cf.* Klenk, 1973). Although these viruses are members of different groups, such as togaviruses, rhabdoviruses, and myxoviruses, their carbohydrate compositions are in general rather similar. The constituent sugars of glycoproteins are galactose, mannose, fucose and glucosamine, those of glycolipids glucose, galactose and galactosamine. However, whereas toga- and rhabdoviruses contain in addition significant amounts of neuraminic acid linked to protein (Burge and Strauss, 1970; Burge and Huang, 1970) and to lipid (Renkonen *et al.*, 1972; Klenk and Choppin, 1971), this carbohydrate could not be detected in myxoviruses (Klenk and Choppin, 1970b; Klenk *et al.*, 1970a). The presence of neuraminic acid in toga- and rhabdoviruses, and its absence in myxoviruses suggests that the enzyme neuraminidase, a component of the latter viruses, is responsible for its absence. Furthermore, it could be shown that the absence of neuraminic acid is confined to those areas of the cell surface where virus maturation is occurring, the rest of the plasma membrane contains neuraminic acid (Klenk *et al.*, 1970b). This suggests that there is a localized action of viral neuraminidase in regions of cell membrane in which this enzyme has been incorporated and which are converted into viral envelopes. It is not clear whether the absence of neuraminic acid on the myxovirus envelope has any impact on the biological behaviour of these viruses. However, it is conceivable that due to its negative charge neuraminic acid as a constituent of the haemagglutinin in glycoproteins of myxoviruses would prevent the attachment of the haemagglutinin to the cellular receptor which is also neuraminic acid. The absence of neuraminic acid would then be essential for the biological function of the haemagglutinin. If this hypothesis is correct the neuraminidase would be a viral enzyme which plays an important role in the biosynthesis of the viral envelope.

SUMMARY

The myxovirus envelope consists of a central lipid bilayer which is coated on its inner surface by a carbohydrate-free protein whereas its outer surface is covered by surface projections (spikes) consisting of glycoproteins. The lipids in the viral envelope are derived from the host membrane and, therefore, show host specificities. However, the envelope polypeptides are specified by the viral genome. The carbohydrates are either bound to lipids or to proteins.

In myxoviruses the glycoproteins of the spikes have haemagglutinin and neuraminidase activities. In the synthesis of the influenza virus haemagglutinin several steps have been identified. First a carbohydrate-free polypeptide is formed which then is glycosylated to a large glycoprotein. This glycoprotein is cleaved into two smaller glycoproteins which form the subunits of the haemagglutinin in the mature virion.

The spike glycoproteins on the surface of the virion contain a carbohydrate receptor specific for Concanavalin A. Thus intact virions can be precipitated by this phytagglutinin. If the spikes are removed, this agglutinability is lost. However the particles can now be precipitated by antibodies and phytagglutinins specific for the carbohydrates of glycolipids which like all other lipids are derived from the host cell. This indicates that in the viral envelope the glycolipids are located in the lipid layer below the glycoproteins of the surface spikes. The serological specificities of the protein-bound and the lipid-bound carbohydrates are different.

The carbohydrate moiety of the viral glycoproteins seems to be specified by the host cell. Although neuraminic acid is a ubiquitous constituent of the plasma membrane of vertebrate cells, no protein-bound nor lipid-bound neuraminic acid was detected in myxoviruses. The lack of this carbohydrate is probably due to the action of the viral neuraminidase.

Acknowledgement

This work has been carried out in collaboration with Dr. P. W. Choppin at the Rockefeller University, New York, and Dr. R. Rott at the University of Giessen, Germany. I am also grateful to Dr. H. J. Eggers for his constant interest and support.

REFERENCES

Bächi, T., Gerhard, W., Lindenmann, J. and Mühlethaler, K. (1969) *J. Virol. 4*, 769–776.

Becht, H., Rott, R. and Klenk, H.-D. (1972) *J. Gen. Virol. 14*, 1.

Burge, B. W. and Huang, A. S. (1970) *J. Virol. 6*, 176–182.

Burge, B. W. and Strauss, I. H., Jr. (1970) *J. Mol. Biol. 47*, 449–466.

Burger, M. M. (1969) *Proc. Natl. Acad. Sci. U.S.A. 62*, 994.

Chen, C., Compans, R. W. and Choppin, P. W. (1971) *J. Gen. Virol. 11*, 53–58.

Choppin, P. W., Klenk, H.-D., Compans, R. W. and Caliguiri, L. A. (1971) *Perspectives in Virology* VII, 127–156.

Compans, R. W. and Dimmock, N. J. (1969) *Virology 39*, 499–515.

Compans, R. W., Holmes, K. V., Dales, S. and Choppin, P. W. (1966) *Virology 30*, 411.

Compans, R. W., Klenk, H.-D., Caliguiri, L. A. and Choppin, P. W. (1970) *Virology 42*, 880–889.

Compans, R. W., Landsberger, F. R., Lenard, I. and Choppin, P. W. (1972) *International Virology 2*, 130–133.

Inbar, M. and Sachs, L. (1969) *Proc. Natl. Acad. Sci. U.S.A. 63*, 1418.

Klenk, H.-D. (1973) *Biological Membranes, 2*, 145–183.

Klenk, H.-D. and Choppin, P. W. (1969a) *Virology 37*, 155–157.

Klenk, H.-D. and Choppin, P. W. (1969b) *Virology 38*, 255–268.

Klenk, H.-D. and Choppin, P. W. (1970a) *Virology 40*, 939–947.

Klenk, H.-D. and Choppin, P. W. (1970b) *Proc. Natl. Acad. Sci. U.S.A. 66*, 57–64.

Klenk, H.-D. and Choppin, P. W. (1971) *J. Virology 7*, 416.

Klenk, H.-D., Caliguiri, L. A. and Choppin, P. W. (1970a) *Virology 42*, 473–481.

Klenk, H.-D., Compans, R. W. and Choppin, P. W. (1970b) *Virology 42*, 1158–1162.

Klenk, H.-D., Rott, R. and Becht, H. (1972a) *Virology 47*, 579–591.

Klenk, H.-D., Scholtissek, C. and Rott, R. (1972b) *Virology, 43*, 723–734.

Landsberger, F. R., Lenard, I., Paxton, I. and Compans, R. W. (1971) *Proc. Natl. Acad. Sci. U.S.A. 68*, 2579–2583.

Laver, W. G. (1971) *Virology 45*, 275–288.

Laver, W. G. and Valentine, R. C. (1969) *Virology 38*, 105–119.

Lazarowitz, S. G., Compans, R. W. and Choppin, P. W. (1971) *Virology 46*, 830–843.

McSharry, I. I., Compans, R. W. and Choppin, P. W. (1971) *J. Virol. 8*, 722–729.

Renkonen, O., Kääräinen, L., Simons, K. and Gahmberg, C. C. (1971) *Virology 46*, 318–326.

Rifkin, D. B. and Compans, R. W. (1971) *Virology 46*, 485–489.

Schulze, I. T. (1970) *Virology 42*, 890–904.

Schulze, I. T. (1972) *Virology 47*, 181–196.

Skehel, I. I. and Schild, G. (1971) *Virology 44*, 396–408.

Stanley, P. and Haslam, E. A. (1971) *Virology 46*, 764–773.

Tiffany, J. H. and Blough, H. A. (1970) *Proc. Natl. Acad. Sci. U.S.A. 65*, 1105–1112.

Uhlenbruck G. (1971) *Chemie, 25*, 10–21.

14 *RNA Tumour Virus-directed Glycoproteins and Their Significance in the Virion and the Cell*

Heinz Bauer,[a] Hans Gelderblom,
Dani P. Bolognesi[b] and Reinhard Kurth[a]
a. Robert Koch-Institut, Abteilung für Virologie, Berlin
b. Duke University Medical Center, Durham, N.C., U.S.A.

It is a widespread belief that some RNA-viruses are causative agents for spontaneous tumours in animals and in man. Such viruses, detected in a wide variety of animals, especially birds and mammals share important biological properties. They differ in host specificity i.e. in hosts where a given virus appears rather than hosts which can be infected experimentally.

A host independent distinction, morphologically gives a classification into A, B or C types (Bernhard, 1960). While A particles are immature forms of B, the latter is distinguished from C by long surface projections, their eccentrically located core and their budding mechanism. The C particles of different species differ serologically and in minor morphological ways. In principle, the structure of a C-type virion is schematically illustrated in Fig. 14.1. This structure was developed on the basis of our morphological, serological and biochemical studies of avian RNA-tumour viruses (ATV) (Bolognesi *et al.*, 1972a; Bolognesi *et al.*, 1972b; Gelderblom *et al.*, 1972). A similar scheme has been proposed by Nermuth *et al.* (1972) after studying the morphology of murine viruses. The core of the virion consists of four proteins (Fig. 14.2) and contains the high molecular weight RNA which is associated with the reverse transcriptase. The envelope represents a "unit"-lipid membrane that is derived from the cell membrane during the budding process (see Fig. 14.7b) and is covered with knob-like projections. The four major core proteins are serologically distinct from each other

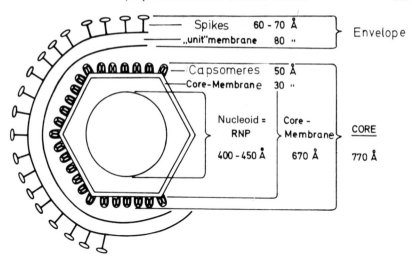

Fig. 14.1 Schematical presentation of the fine-structure of ATV. The total diameter of the particle is 1000 to 1200 Å.

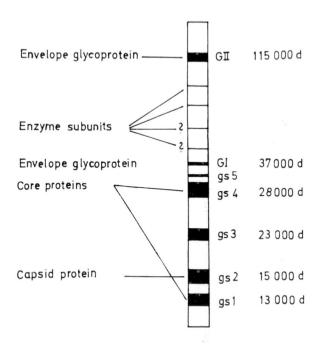

Fig. 14.2 Analysis of the major protein constituents of ATV by SDS-polyacrylamide gel electrophoresis.

(Bauer and Bolognesi, 1970; Fleissner, 1971) and at least in part, group-specific (gs) for the ATV (Bauer, 1970). In contrast, the antigenicity of the envelope antigen (Ve) is characteristic for each virus strain and served as the basis for the classification of these viruses into the subgroups A through E (Vogt, 1970; Hanafusa et al., 1970), where viruses of the same subgroup have related Ve antigens.

VIRUS-ENVELOPE ANTIGENS

Antigens play an important role and attract much attention in tumour virus research because they not only serve as markers for the detection and assaying of the virus, but must also be considered as molecules with a possible biological function in the transformation process. The envelope antigen is of special interest because it can be used as a vaccine. Although the first isolation of a relatively pure internal gs antigen was described in 1965 (Bauer and Schäfer, 1965), it was not until 1970 that we could report the first isolation of a type-specific antigen (Tozawa et al., 1970). For that purpose, methods different from those applied successfully to other enveloped viruses had to be developed, probably because of a higher lipid content of the oncornaviruses. Finally, after treatment of the virus with Tween 20, a material was released that possessed all the properties attributed (Vogt et al., 1967) to the type-specific envelope antigen of avian RNA tumour viruses. It absorbed, induced the synthesis of, and was precipitated (Fig. 14.3) by type-specific neutralizing antibody; furthermore, it blocked the genetically determined (Crittenden et al., 1963) cell receptors specific for viruses of the same subgroup (Tozawa et al., 1970).

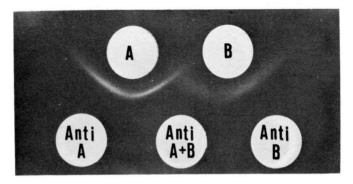

Fig. 14.3 Immuno-precipitation of Ve antigen. A: Ve antigen of subgroup A virus. B: Ve antigen of subgroup B virus. Anti A and Anti B resp.: neutralizing chicken serum against virus of subgroup A and B resp. Anti A + B: mixture of the two antisera.

By phenol-SDS extraction, the Ve antigen could also be recovered, although in an antigenically less active form. After isolation by preparative polyacrylamide gel (PAG) electrophoresis, it was shown by carbohydrate staining as well as by radioactive labelling to contain glycoprotein migrating at a position in the PAG gel that corresponds to a molecular weight of 115,000 daltons when compared with standard proteins (Bolognesi and Bauer, 1970; Bauer and Bolognesi, 1970).

The procedure was much improved recently in that we now isolate the Ve antigen by ultracentrifugation in a sucrose gradient after Nonidet-P40 treatment of the virus (Bolognesi *et al.*, 1972a). This results in almost quantitative yield of the antigen being pure and antigenically active in a form of rosettes (Fig. 14.4a). From the

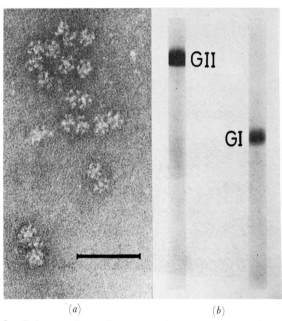

(a)   (b)

Fig. 14.4 Morphology and constituents of Ve antigen. (*a*): Isolate of rosette-like aggregates of the knob-like virus surface projections (Bar represents 1000 Å). (*b*): Analysis of the two separated rosette subunits by protein staining after PAG-electrophoresis. GII represents the knob with a molecular weight of 115,000 dalton and GI the spike of 37,000 dalton.

dimensions of the rosette subunits, it is obvious that they represent aggregates of the knob-like surface projections of the virion. Biochemical analysis of these structures reveals that they consist of two glycoproteins of 37,000 daltons (GI) and 115,000 daltons (GII) (see Fig. 14.2) respectively, with the former having a much lower carbohydrate content. It was evidenced that the 115,000 dalton molecule associates with the knob. As a consequence we assumed that the 37,000 dalton molecule is located in the spike of the total virus surface projection.

The two glycoprotein constituents could be separated by velocity ultracentrifugation in a sucrose gradient after sodium dodecyl sulphate (SDS)-treatment (Fig. 14.4*b*). This permitted the serological analysis of GI and GII which first revealed that both contained type-specific antigenicity. Both not only induced early interference by blocking the cell receptors, but could also be precipitated by neutralizing chicken antibody. Since their Ve antigenicity was highly decreased by SDS, this was only possible by co-precipitation of radioactively-labelled Ve antigen with intact rosettes (Fig. 14.5*a, b*). The resolution of that immuno-precipitation,

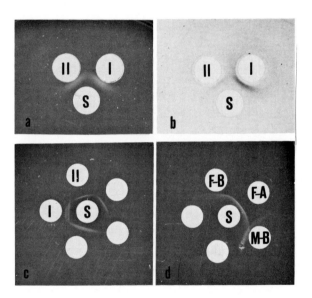

Fig. 14.5 Serological analysis of GI and GII. *a, b*: immuno-coprecipitation by chicken antiserum (S) of radioactively labelled GII (II) and GI (I) with unlabelled rosettes; photography (*a*) and autoradiography (*b*) of the precipitate. *c*: demonstration of non-type-specific antigens in GI (I) and GII (II) by precipitation with a non-neutralizing rabbit serum (S). *d*: demonstration of the host cell specificity of an antigenic site in GI and GII. Non-neutralizing rabbit serum (S) is reacted with rosettes from subgroup A and B virus grown in fibroblasts (F–A and F–B resp.) and from subgroup B virus grown in myeloblasts (M–B).

however, is not such to permit the decision whether GI and GII are of the same or different antigenicity. Since the cross-neutralization within one subgroup is not complete it could well be that one constituent represents the type- and the other a subgroup-specific antigenicity. We hypothesized this earlier to explain a partial cross-reaction between viruses of subgroup B and D (Bauer and Graf, 1969).

In screening experiments, we found that some gs rabbit sera that had been prepared against SDS-extracted viral antigens (Eckert *et al.*, 1964; Bauer and Schäfer, 1965) and were lacking neutralizing capacity, precipitated the rosettes as well as the two separated subunits GI and GII. Furthermore, the involved antigens were not identical with the Ve antigen (Bolognesi *et al.*, 1972a) and also different in GI and GII (Fig. 14.5c). It is assumed that the chemical substrate of these antigens is the carbohydrate part of the glycoproteins which probably is cell-specific. We carried out the following experiment in support of this assumption: When total Ve antigen from a subgroup A (Schmidt-Ruppin Rous sarcoma virus (SRV) A) and subgroup B (myeloblastosis-associated virus (MAV) B) virus were reacted with the rabbit serum in comparison with avian myeloblastosis virus of subgroup B (AMV-B), the SRV-A and MAV-B reacted identical by but different from AMV-B (Fig. 14.5d), although MAV-B and AMV-B share a type-specific Ve antigen. Since SRV-A and MAV-B are both synthesized in chicken embryonic (mostly fibroblast) cells and AMV-B in bone-marrow-derived myeloblast cells, we assume from this experiment that Ve antigens even of different subgroups are glycosylated similarly in a given cell type, but different in heterogeneous cell types. Also, this carbohydrate part defines a second antigen in the glycoprotein projections of the virus envelope. From this we conclude that the polypeptides of the glycoproteins represent the virus type- or subgroup-specific antigens. That these are probably coded for by the virus genome is supported by recent studies on the messenger function of the viral RNA in an *in vitro* protein synthesis system. The analysis in the PAG-electrophoresis of the protein synthesized *in vitro* with viral 62S RNA as messenger revealed several peaks coinciding with the gs-antigens of the virion and two further peaks that have not yet been examined serologically. According to their molecular weight, they could very well represent the protein parts of GI and GII (Siegert *et al.*, 1972).

VIRUS-DIRECTED CELL SURFACE ANTIGENS

It has been known for several years that a virus-induced cell transformation is accompanied by the appearance of new cell surface antigens. They are characteristic for a given virus and may provoke an immunological response in the tumour bearing host. In the case of ATV it was believed for a long time that the Ve antigen of the virion was the only one possessing this function, although there was some evidence against this assumption. For example,

Rous sarcoma virus (RSV)-induced mammalian sarcomata contained a new tumour-specific cell surface antigen (TSSA) while lacking synthesis of virus particles (Gelderblom *et al.*, 1970). In addition, no virus-neutralizing antibodies could be isolated from the animals (Koldovsky *et al.*, 1966; Bauer *et al.*, 1969).

We investigated the nature of ATV directed cell surface antigens by use of different seroimmunological techniques, namely immunoferritin and fluorescent antibody staining of cells, as well as cytotoxic assays with immune lymphocytes or antisera (Gelderblom *et al.*, 1972; Kurth and Bauer, 1972a). The target cells were chicken embryonic cells (CEC) either productively infected by an avian leukosis virus (ALV) or transformed by a sarcoma virus (ASV). Immune chicken sera taken after a single virus injection and immune chicken lymphocytes obtained after repeated virus injection were used.

The results indicated that not only the membrane of virus, but also the surface of both transformed and productively infected CEC were stained by ferritin-labelled antibody when directed against virus of the homologous subgroup to which infecting virus belonged (Fig. 14.6). Cells infected by ALV of subgroup B, on the other hand, were not stained by sera against subgroup A and vice versa. We concluded that the relevant antigen is identical with the type and subgroup specific Ve antigens. It is interesting to note that in general, virus budding took place in Ve antigen-free cell surface areas, although the Ve antigen was demonstrable in patches of 0.5 to 2 μ in size without a sharp delineation.

Also demonstrated was a second kind of antigen which was restricted to ASV-transformed cells and detectable only by ASV-antisera. This reaction was not subgroup but rather group-specific. Antisera prepared against a given sarcoma virus stained ASV transformed CEC regardless of the group of the virus used for transformation. Thus, SRV-H (subgroup D) transformed cells, though not stained by ALV-A antisera, were stained by SRV-A antiserum and vice versa (Fig. 14.7). This antigen is not identical with either of the major protein constituents of the virion and is called TSSA (tumour specific surface antigen). Its close correlation with the transformed state of a cell was also demonstrated by its absence on cells that were infected by non-converting-mutants derived from ASV. They had lost the capability to transform CEC *in vitro* but still replicated in those cells (Graf *et al.*, 1971).

It is remarkable that appropriate immune lymphocytes exerted a subgroup specific cytotoxic effect not only via the TSSA but also via the Ve antigen, though to a weaker extent. Productively infected cells lacking the TSSA were also killed in a cytotoxic assay by appropriate immune lymphocytes (Kurth and Bauer, 1972a).

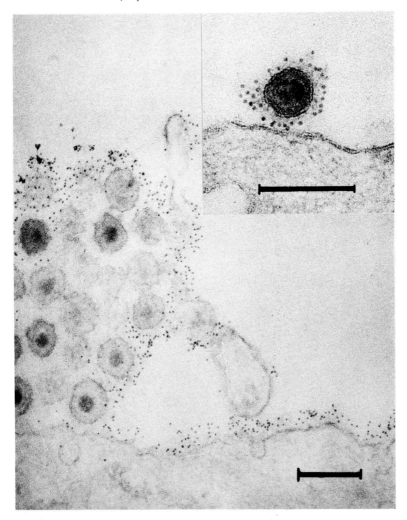

Fig. 14.6   Demonstration of Ve antigen by the immuno-ferritin technique on the surface of the virion as well as of the cell. Bar represents 2000 Å.

There was one discrepancy between the results obtained with the immuno-ferritin and the cytotoxic tests. While ALV-antisera did not detect TSSA, lymphocytes from chickens infected with ALV did so. We believe that this is due to the different immunization technique. Immune sera were taken after a single virus injection, while immune lymphocytes were taken after a series of booster injections, that is several months after the first virus injection.

This longer period for tumour-induction in ALV (Burmester *et al.*, 1960) is accompanied by development of enough TSSA to immunize the chicken. This confirms previous results (Bauer *et al.*,

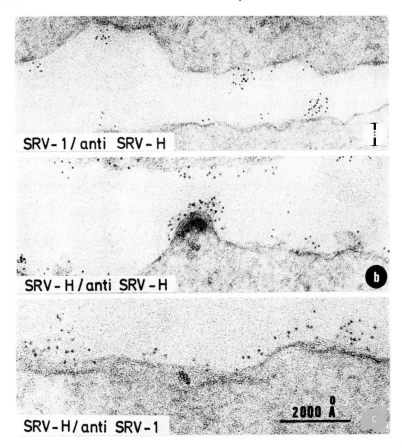

Fig. 14.7 Demonstration of the group specific TSSA by the immuno-ferritin technique, *a*: SRV-1 (subgroup A) transformed cell stained with SRV-H (subgroup D) antiserum, *b*: SRV-H transformed cell and virus particle stained with homologous antiserum, *c*: SRV-H transformed cell stained with SRV-1 antiserum.

1969) that there is a common TSSA in different target cells transformed by viruses of the same group.

The origin of ATV induced TSSA is completely unknown. Even considering the virus specificity, in that TSSA induced by viruses of different groups do not cross-react one has still to entertain the possibility that these antigens might be cell coded and, while not expressed in normal cells in a detectable form, they are induced during the transformation process. In that connection then, it would be interesting to know whether the ATV specific TSSA induced in mammalian cells and the TSSA induced in natural host cells are antigenically related. In the original detection of ATV-induced TSSA in mammals, Sjögren and Jonsson (1963) described a

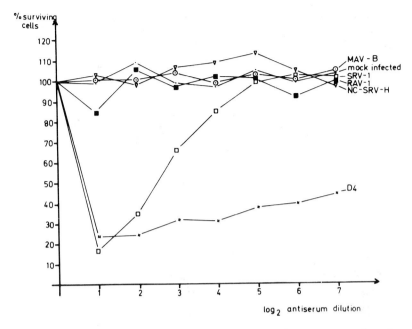

Fig. 14.8 Titration of the cytotoxic effect of mouse antiserum prepared against isogeneic ASV tumour cells that is reacted with isogeneic mouse tumour cells (d4) and CEC. Only transformed CEC (SRV-1) were damaged, but not productively infected (MAV-B, RAV-1, NC-SRV-H) or mock infected CEC.

"resistance against isotransplantation of mouse tumours induced by Rous sarcoma virus". Later on we demonstrated that these tumour specific transplantation antigens in mice are group specific for different ASV strains (Bubenik and Bauer, 1967) and even cross react between leukosis and sarcoma virus (Bauer *et al.*, 1969). In order to study the relationship between ATV induced TSSA in different species we undertook a series of experiments comparing ATV transformed cells of different species using the *in vitro* assays for TSSA as described above. Cells were from chicken, mice and hamster, immune sera from chicken and mice and immune lymphocytes from chicken. It turned out that there was a distinct immunological cross reaction between the TSSA of these cells (Kurth and Bauer, 1972b; Gelderblom and Bauer, 1973). As an example, serum from mice hyperimmune against isogeneic Rous sarcoma cells (Kurth and Bauer, 1973) exerted a complement-dependent cytotoxic effect not only against homologous mouse tumour cells, but also against ASV transformed chicken cells (Fig. 14.8) and hamster cells (not shown in Fig. 14.8). As a control normal and ALV-infected chicken cells were not killed (Kurth and Bauer, 1972b). On the other hand, ASV-mouse tumour cells were stained by ferritin labelled chicken

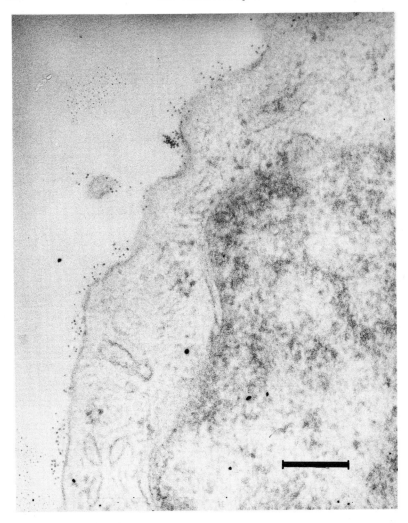

Fig. 14.9   Detection of common TSSA on ASV-transformed mouse cells by chicken
ASV-antiserum. Bar represents 2000 Å.

immune sera against ASV (Fig. 14.9) and damaged by immune
chicken lymphocyes (Table 14.1). Since ALV-chicken antisera did
not react with ASV-transformed mammalian cells and mouse antiserum did not stain chicken cells infected with an ALV of the
subgroup of the mouse cell transforming ASV, we believe that the
cross reacting TSSA in ATV transformed cells of different species is
not the Ve antigen.

   These data might support the general hypothesis that virus
induced TSSA are coded for by the virus; however, they are not

Table 14.1
SURVIVAL OF AVIAN SARCOMA VIRUS TRANSFORMED
MOUSE CELLS AFTER INCUBATION WITH IMMUNE
LYMPHOCYTES FROM CHICKEN

Effector cells taken from chicken immunized with the following:

Not immunized	ALV of subgroup A (RAV-1)	Non-converting mutant of subgroup A (NC-SRV-A)	SRV-A	B 77 (subgroup C)	SRV-D
71.0 ± 7.5	50.3 ± 7.5	33.1 ± 6.4	28.4 ± 6.2	19.9 ± 7.7	16.6 ± 3.1

Figures represent percentage of surviving mouse tumour cells after incubation in a ratio of 1:25 with chicken spleen cells compared to the growth of target cells in medium only.

conclusive in that respect. It is very possible that a virus induces cellular antigens of related antigenicity in cells of different species, such as antigens specific for the embryonic state of a cell, of which countless numbers might exist. At the present, experiments do not allow the appearance of new antigens to be correlated with changes observed in the membrane of transformed cells. By analogy with natural transplantation antigens (Muramatsu and Nathenson, 1970) one may assume them to be glycoprotein in nature.

SUMMARY

The envelope of avian RNA tumour viruses is covered with projections which represent the type specific virus envelope antigens and consist of two individual glycoproteins GI and GII. Evidence was presented indicating that the protein part of both GI and GII represents the type and/or subgroup specific viral antigens that induce neutralizing antibody and have specific affinity to the cell receptors for virus infection. The carbohydrate parts of GI and GII are different from each other not only in size. They define a second antigenic specificity and are thought to be host-cell derived. The Ve antigen is demonstrable on the surface of both productively infected and transformed chicken cells but not in non-productively transformed mammalian cells. Though to a lesser extent, it acts as a transplantation antigen similar to a further virus specified cell surface antigen (TSSA) that is restricted to transformed cells, group specific for that virus group, and antigenically related in ATV transformed cells of different species. The origin and nature of these TSSA are unknown.

REFERENCES
Bauer, H. (1970) Zbl. f. Vet. Med., B 17, 582.
Bauer, H. und Schäfer, W. (1965) Z. Naturf. 20b, 815.

Bauer, H. and Graf, Th. (1969) *Virology 37*, 157.

Bauer, H. and Bolognesi, D. P. (1970) *Virology 42*, 1113.

Bauer, H., Bubeník, J., Graf, Th. and Allgaier, Ch. (1969) *Virology 39*, 482.

Bernhard, W. (1960) *Cancer Res.*, *20*, 712.

Bolognesi, D. P. and Bauer, H. (1970) *Virology 42*, 1097.

Bolognesi, D. P., Bauer, H., Gelderblom, H. and Hüper, G. (1972a) *Virology 47*, 551.

Bolognesi, D. P., Gelderblom, H., Bauer, H., Mölling, K. and Hüper, G. (1972b) *Virology 47*, 567.

Bubeník, J. and Bauer, H. (1967) *Virology 31*, 489–497.

Burmester, B. R., Fantes, A. K., Waters, N. F., Bryan, W. R. and Groupé, V. (1960) *Poultry Science 39*, 199.

Crittenden, L. B., Okazaki, W. and Reamer, R. H. (1963) *Virology 20*, 541–544.

Eckert, E. A., Rott, R. and Schäfer, W. (1964) *Virology 24*, 426–433.

Fleissner, E. (1971) *J. Virol. 8*, 778–785.

Gelderblom, H., Bauer, H. and Frank, H. (1970) *J. gen. Virology 7*, 33.

Gelderblom, H., Bauer, H., Bolognesi, D. P. und Frank, H. (1972) *Zbl. Bakter. Hyg. I. Abt. Orig. A 220*, 79.

Gelderblom, H., Bauer, H. and Graf, Th. (1972a) *Virology 47*, 416.

Gelderblom, H. and Bauer, H. (1973b) *Internat. J. Cancer*, *11*, 466.

Graf, Th., Bauer, H., Gelderblom, H. and Bolognesi, D. P. (1971) *Virology 43*, 427–441.

Hanafusa, H., Hanafusa, T. and Miyamoto, T. (1970) *Proc. Natl. Acad. Sci. U.S.A.*, *67*, 1797–1803.

Koldovsky, P., Svoboda, J. and Bubeník, J. (1966) *Folia Biol. (Prague) 12*, 1–10.

Kurth, R. and Bauer, H. (1972a) *Virology 47*, 426–433.

Kurth, R. and Bauer, H. (1972b) *Virology 49*, 145–159.

Kurth, R. and Bauer, H. (1973) *Eur. J. Immunol.*, *3*, 95.

Muramatsu, T. and Nathenson, St. G. (1970) *Biochem. Biophys. Res. Comm. 38*, 1–8.

Nermuth, M. V., Frank, H. and Schäfer, W. (1972) *Virology* (in press).

Siegert, W., Konings, R. N. H., Hofschneider, P. H. and Bauer, H. (1972) *Proc. Natl. Acad. Sci. U.S.A.*, *68*, 888.

Sjögren, H. O. and Jonsson, N. (1963) *Exptl. Cell. Res. 32*, 618.

Tozawa, H., Bauer, H., Graf, Th. and Gelderblom, H. (1970) *Virology 40*, 530.

Vogt, P. K. (1965) *Virology 25*, 237.

Vogt, P. K., Ishizaki, R. and Duff, R. (1967) "*Subviral Carcinogenesis*" (Y. Ito, ed.) p. 297–310.

Index